Quick & Basic
Troubleshooting

*A Contractor's Easy Guide to
Fixing HVAC Wiring & Circuits*

Quick & Basic Troubleshooting

Carol Fey

P.I.G. Press
Littleton, Colorado

Disclaimer: This manual is intended as a tool for a classroom setting. Do not attempt any of the instructions in this book without the guidance of a qualified instructor.

Copyright © 2001 by Carol Fey

All rights reserved. No part of this book may be reproduced or transmitted in any form or by any means, electronic or mechanical, including photocopying, recording, or by any information storage and retrieval system, without permission in writing from the publisher.

Published by P.I.G. Press
759 E. Phillips Drive S.
Littleton, CO 80122-2873

ISBN No.: 0-9672564-2-9

Library of Congress Control Number: 2001118693

Printed in the United States of America

This is dedicated to my fellow industry educators, in particular my technical support team (in alphabetical order) Bob Boltz, Mark Eatherton, John Edwards, Dan Holohan, Alan Levi, Roger Schilling, and Dale Watterson.

Contents

Introduction 11

Review of the basic circuit 21

Troubleshooting with a meter 26

Troubleshooting individual components 36
    Troubleshooting power supplies 36
    Troubleshooting switches 41
    Troubleshooting loads 44
    Troubleshooting relays 45
    Troubleshooting panels 46

Troubleshooting a complete circuit 48
    Using "home-run," "hop-scotch," "leap frog," or
    "daisy chain" methods 48
    Checking your wiring 59
    Using a troubleshooting chart 62
    Using a ladder diagram 80

QUICK & BASIC TROUBLESHOOTING

Final exam 92

Index 93

Order information 95

Acknowledgments

Thank you to Honeywell for providing the basic concepts for many of the drawings included in this book.

About the Author

Carol Fey is a degreed technical trainer who has worked more than 20 years in the controls industry. She has been honored as National Technical Trainer of the Year by the American Society for Training and Development. She is also the author of *Quick & Basic Electricity: A Contractor's Easy Guide to HVAC Circuits, Controls, and Wiring Diagrams,* and *Quick & Basic Hydronic Controls: A Contractor's Easy Guide to Hydronic Controls, Wiring, and Wiring Diagrams.*

Introduction

I spent the summer in my basement. It wasn't the coolest spot—actually the rest of my house was colder. That was the problem. I was troubleshooting my air conditioning system. Where I live, in Colorado, the HVAC equipment is in the basement. It took all summer because I was doing it wrong.

I was doing it wrong just like you might. I didn't follow a troubleshooting process. I skipped all around. I made assumptions. I replaced parts. I took advice.

Troubleshooting is a process. You have to start at the beginning and do all the steps without skipping any. If you do that, you'll find out what's wrong. If you don't, you'll just waste a lot of time.

Here's how I set myself up to spend my summer troubleshooting in the basement. See if it sounds familiar. You wet-head hydronics folks are going to tempted to say "serves ya right" for having forced air, but hey, it came with the house.

QUICK & BASIC TROUBLESHOOTING

I had added forced air zoning to my heating and air conditioning system. Forced air zoning is a great system. Like hydronic zoning, it makes it possible to deliver different amounts of heat (or cooling) to different parts of the house. Don't blame my troubleshooting problems on zoning. Blame them on my not following the troubleshooting process.

My zoning system worked fine at first. There are six zones. Any one can call for heating or cooling whenever it needs it. The dampers for that zone open, while all the others close. The dampers and thermostats for those six zones are wired into two electronic zoning panels, three zones on each. The panels are wired together so that they think as one brain.

And then there was a lightning storm. After the storm, when any one of the first three zones called for cooling, that zone plus the last three zones cooled. When any of the last three zones called for cooling, nothing at all happened. Aaagh.

I called the 800-number technical hotline. They really are great guys. They had me check lots of voltages with my meter. Everything checked out. "Well, bad news," they concluded. "You need to replace the number one panel." I remarked, with a big sigh and a chuckle, "So it took us an hour to get where a lot of contractors just start—change out the electronic panel!"

Here's a routine you'll recognize. I drove to the supply house, picked up a replacement panel, and killed a couple more hours talking to people I know. I got back to the job, *labeled all those wires on the panel* with masking tape, removed *all those wires* from the first panel, re-wired *all those*

Introduction

wires to the replacement panel, crossed my fingers, turned the power on, and *voila*——nothing had changed. Hum.

So I removed all the wires from the replacement panel, re-wired them back to the original panel, and put the replacement panel back in the box. That about killed the day.

Next day, back to the hotline. "Hummmm. Let's check those voltages again. Yep. If it's not the first panel it's just gotta be the second panel. Replace it."

Sooo, bigger sigh this time, because I *really* knew how much time this was going to take. But if it's what I have to do . . . So I drove to the supplier again, but they didn't have that panel. I drove to a second supplier, got the panel, took it to the job, labeled all the wires on the second panel, unwired the second panel, rewired onto the replacement panel, crossed my fingers, held my breath, turned on the power, and . . . nothing had changed.

You know what I did next and how I felt about it. Another day killed.

By now I'd completely forgotten everything I may have ever known about good troubleshooting process. I started asking for advice. I asked everybody, and I got a lot of sympathy. The consensus was to change the transformers.

I got two new transformers. I remembered the rule to never work alone with line voltage if at all possible, so I reviewed with my teenaged daughter what to do should she find me electrocuted (here's the wooden broom handle—touch me only with this, and call 911). She threatened to leave home until I got my sanity back.

QUICK & BASIC TROUBLESHOOTING

I shut off the basement circuit breakers, went to the basement, found everything dark, went back up and turned the circuits back on, and returned to the basement to find the trouble light and a long extension cord. I plugged the cord into a non-basement circuit, shut off the basement circuit breakers again, and followed my shaky trouble light back to the basement. This stuff takes forever!

Down at last to the task of changing the transformers, I labeled wires, changed the transformers, and rewired them to the panels. I turned on the circuit breakers, applied power to the panels, held my breath, looked heavenward, and . . . (ta-dah!)—nothing had changed.

Out of ideas and advice, I continued to control my cooling by blocking registers with piles of books.

Many weeks later I was having a phone conversation with my troubleshooting zoning-guru friend John. We were commiserating about how really strange this problem is. And we realized—finally—that *I didn't start at the beginning of the troubleshooting process.*

So I got on my cordless telephone headset with John on the other end of the line, and here's what we asked:

- Is there a call for cooling? (Is the thermostat switch closed/turned on?) Yes.
- Is there at least 24volts (V) coming into each of the two panels from the transformers? Yes.
- Is the system working properly? No.

Introduction

- Do the panels work properly with nothing attached to them? Don't know.
- *Don't know?!!!!!*
- You mean I did all this changing of parts and I don't know if the panels work?!!
- Ok. One more time I made sure all the wires were labeled, and removed all of them from the panels. (Oh no, oh no, oh yes.)
- I put 24V back on the panels. I checked to make sure it's still 24V. It is.
- The LEDs on the panels showed that the panels are powered.

[Interesting but non-essential information. LED means *light-emitting diode*. In electronics a diode is a component that makes sure electricity can go only one way in a circuit. A light-emitting diode makes light as well.]

- I connected the zone 1 thermostat to the panel.
- I connected the zone 1 motor to the panel.
- I asked, on a call for cooling from the thermostat, does the motor activate? Yes.
- Let's pause for a minute here. The problem all summer was that when any one of the first three zones (all on the first panel) called for cooling, cooling was also delivered to all three zones on the second panel, even if they didn't want it.

- I connected the first panel to the second panel, and called for cooling for a zone connected to the first panel, in order to see what happened on the second panel. Nothing. That's good.
- So I had new and important information. When everything was removed from the panels except power, the zone 1 stat, and the zone 1 motor, both panels worked properly. This told me that *I didn't have a bad panel or transformer.* (Wish I'd known that at the beginning of the summer!)
- I added the zone 2 stat. Both panels still worked.
- I added the zone 2 motor. The LEDs on the second panel went crazy—just like they'd been doing all summer. They flashed green and red and even kinda pink. Then they settled down to showing cooling for all three of those zones.

So the problem was isolated. There was something about the zone 2 motor that made the second panel go nuts.

The problem had to be that zone 2 motor. Luckily I had a spare. John got off the phone, and I ran to the garage to get my spare motor.

It was about 10 p.m.—hardly the ideal time for working with tiny screws in tight spaces. But you know that. Just about the time I dropped the second screw from the old motor, I noticed a damaged spot on the insulation of the wire coming into the motor. It looked like the wire might be shorted. This was in a place that I *never* would have seen if I hadn't been trying to get positioned to remove that motor.

Could that be it?!! Was that little piece of damaged wire the cause of a whole summer's worth of problems?

You can imagine how I wished I'd seen that damaged spot before I removed (and lost) those two motor screws.

To end a long story quickly, I cut out the shorted section of the wire and just for good measure replaced the motor. I held my breath, crossed my fingers, turned on the power, created a call for cooling on the first zone, and *voila!* the system worked perfectly.

Whatta shame—all that parts changing and wiring and rewiring, all because I didn't find—or even look for—a damaged wire.

The Importance of a Process

We hate following a process. We want freedom! Yet, there are many essential processes that we are willing to follow. Starting a car (or a truck for you *real* men) comes to mind. There is an exact order to starting your car. You start at the beginning—it's pretty much put the key in the ignition and turn it. If you can't find your key (a little like "I can't find my meter" in troubleshooting), you know you can't just skip that step. And you wouldn't listen to a buddy who said, "Ah heck, you don't need that key—just get in and drive."

QUICK & BASIC TROUBLESHOOTING

In troubleshooting, because we don't know there is a process, or we don't remember what the process is, we tend to start in the middle. And we tend to skip around. As you're tempted, remember how far you'd get down the road in your car without following the car-starting procedure.

Types of Troubleshooting

There are several ways to troubleshoot. You can test individual components. You can test the whole circuit. You can use a troubleshooting chart. You can use the "home-run" or "hop-scotch" method. You can use a ladder diagram. We will cover all of these in this book. But, first, before we being actual troubleshooting methods. . . .

What You Need Before You Troubleshoot

Before you begin to troubleshoot, you need the right *attitude*. The right attitude consists of these beliefs:

- This might take awhile. I'm going to work slowly and carefully—agony for those of us with the get-it-done-fast mentality. Imagine yourself as the thoughtful old guy gently rubbing his chin as he just kinda mulls things over . . .
- There is a correct order for troubleshooting, and I'm going to follow it.
- I will not skip steps, I will not skip steps, I will not skip steps.
- I will not jump to conclusions.
- I will not believe what my buddies tell me.

Introduction

- I will not troubleshoot when I'm tired, upset, angry, or hungry (I know, so when else is there).

You need the environment to be safe and reasonably comfortable. Minimum is

- Adequate light
- No water on the floor
- Someone nearby in case you need emergency help

You need some basic *tools* for troubleshooting (details follow in the book)

- A multimeter, preferably digital, with the functions volts AC (VAC), volts DC (VDC), continuity or resistance/ohms, and amps AC
- For safety, a voltage stick, which lights up near 120 volts
- A package of alligator clip jumper wires
- To protect the transformer, a 3-amp fuse link on the transformer secondary (24V) side, or a reset-able micro breaker on the transformer primary (120V) rated for 1.5 to 2 amps
- Screwdrivers—one slotted and one Phillips head, or a combination model
- Needle nose pliers, or angled needle nose pliers
- A flashlight to use very temporarily to locate the problem—use a trouble light for actually doing your work

QUICK & BASIC TROUBLESHOOTING

- A trouble light with a place to plug it in and a place to hang it

Here's what you have to *know* before you can troubleshoot

- The concept an electrical circuit
- How to use a meter
- How to read a troubleshooting chart
- How to read a ladder diagram
- Sequence of operation of system or device you're troubleshooting

If you don't know these things, I suggest you read the book *Quick & Basic Electricity: A Contractor's Easy Guide to HVAC Circuits, Controls, and Wiring Diagrams*. Here's a quick review of the gotta-know's.

Review of the Basic Circuit

Circuit rule number one is that there must be at least one of these three things: a power supply, a switch, and a load.

We'll talk about troubleshooting each of these three later on. Meanwhile,

- A power supply is either line voltage (120V) as you would find in any household electrical outlet, or 240V, or 24V from a transformer.
- A switch turns the circuit on and off. Thermostats and limits are the most common switches.
- A load uses electricity to get a job done. Valves are common loads. They have a motor which uses electricity to open and close the valve.

Circuit rule number two is that these three things have be connected by wires to make a complete circle.

QUICK & BASIC TROUBLESHOOTING

This means you have to think in circles.

When thinking about controls, think about three. When thinking about circuits, think in circles. Say to yourself—power supply, switch, load, and back to power supply—complete circle. Power supply, switch, load, and back to power supply—complete circuit. It's three in a circle.

Imagine the power supply, switch and load all joining their little imaginary wire-like hands for ring-around-the-rosie. As long as they stay joined together, things can happen. But if there's a break in the circle, everything stops.

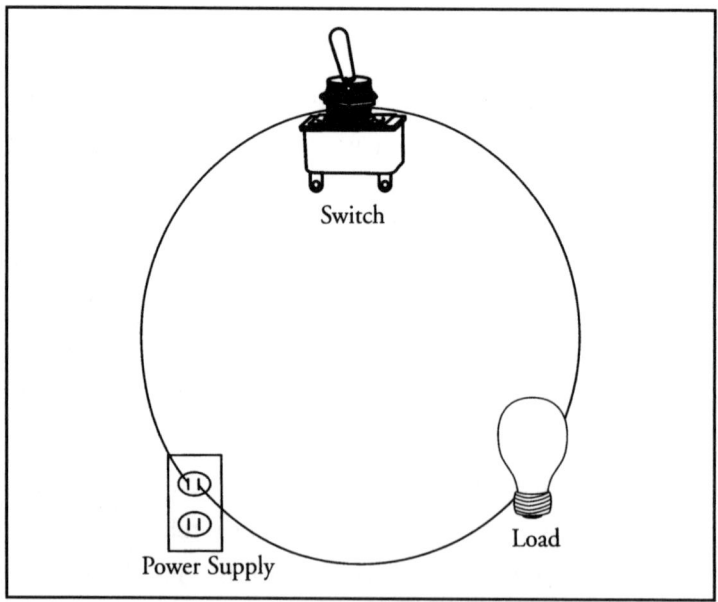

Figure 1. A circuit as a circle

A tricky thing about circuits is that they don't *look* like circles. In a drawing or diagram, they look like squares or rectangles. They have corners.

Review of the Basic Circuit

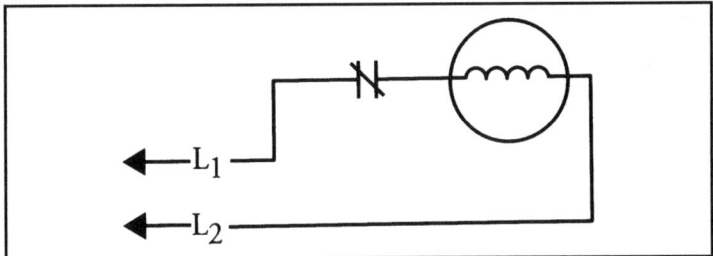

Figure 2. A circuit as a rectangle

In real life, a circuit often looks more like a pile of spaghetti. But you can do some easy translating. Think like a bug! Picture a rope with the ends tied together—it's continuous. Lay it on the floor. Now imagine you're a bug. If you take off along the rope in one direction, you'll eventually get back to where you started—no matter what shape the rope's in.

Figure 3. Test for a complete circuit by thinking of a bug on a rope

The rope can be a circle or rectangle or even a pile of spaghetti. From the your bug's perspective it's a complete circuit as long as you can get back to where you started. It doesn't matter where or which direction you start.

If there are knots in the rope, the bug can go over them as if they didn't exist. The same goes for electrical connectors. A tight wire nut or other electrical connector is like a knot in a rope.

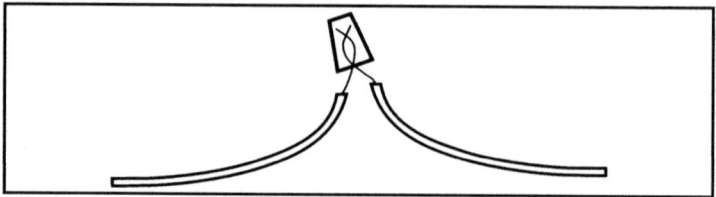

Figure 4. A wire connector

Just as it doesn't matter to the bug if the rope is a different color in various places, it doesn't matter to electricity what color a wire is.

One big difference for electricity, though, is that the path has to conduct electricity. In our business, that usually means metal.

Circuits can be wired in *series* or in *parallel*. This terminology refers to how loads are put together. Let's take a quick look at what each means. For more detail, you might look at the book *Quick & Basic Electricity*.

Here is a series circuit.

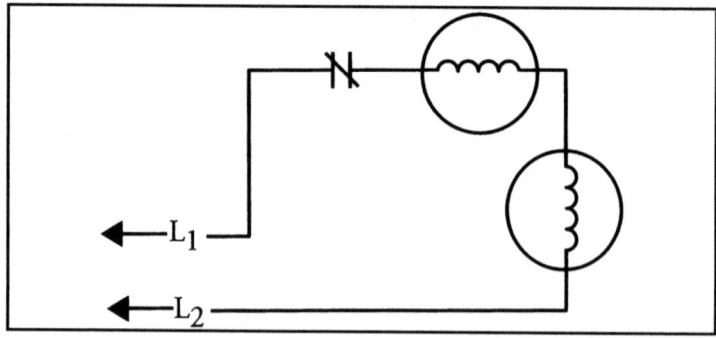

Figure 5. A series circuit

Review of the Basic Circuit

In a series circuit loads are strung together like beads on a necklace. In HVAC work, we rarely put loads in series. The classic example of loads in series is old-fashioned Christmas tree lights. If one goes out, they all go out.

A much more practical way to wire loads is called parallel. You can still wire several loads and their switches to one power supply. But each load has its own private *access* to the power supply, even though it's not the only load. In a string of lights, this means that if one goes out, the others keep on working.

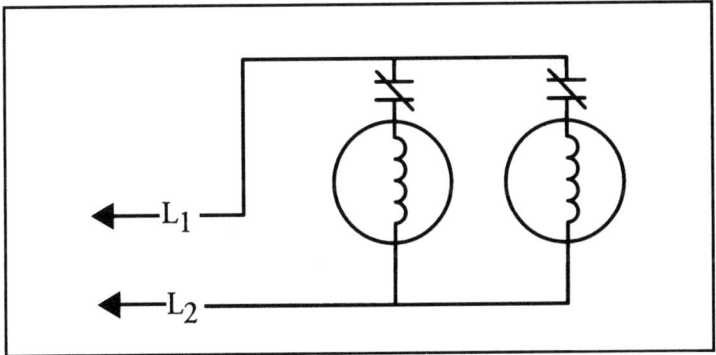

Figure 6. A parallel circuit

These loads are *in parallel* with each other (notice how the loads are on parallel lines). Strictly speaking, they are all in one parallel circuit. That's because they all have the same power supply.

Troubleshooting with a Meter

Not having a meter is like driving with your eyes closed. You can bumble and grope around, but sooner or later you're going to crash.

Buying a Meter

A meter doesn't have to be expensive. But paying more generally means you get more, and it will probably last longer. Here are some things to consider.

- If you're an HVAC professional, go to your favorite HVAC distributor and buy an electronic multimeter. It's a business investment, so a little expense is really OK.
- It's called a multimeter because it has multiple functions. The functions you need are:
AC voltage (called VAC) for alternating current circuits
DC voltage (called VDC)—for batteries and direct current circuits

Troubleshooting with a Meter

    AC amperage (often not available on cheaper meters)
    Continuity (sometimes known as resistance, or ohms)

[Interesting but non-essential information. Most HVAC control circuits are AC. You'll use your DC meter function mostly to check batteries. DC circuits are mostly in electronics and large direct digital control (DDC) systems.]

When you buy an *electronic* meter, you get a digital read-out. That's good because it's easy to read. An electronic meter measures in decimals (numbers to the right side of the dot). Most of the time you can pay attention to just the whole number (the digits to the left of the dot).

Electronic meters are fairly durable. They can usually withstand a little bumping around—maybe even being dropped. That alone is a good reason to buy the electronic meter.

If your level of investment has to be in the $10 range, there's something for you. You can buy a cheap meter at a neighborhood electronics store. It'll be plenty accurate for troubleshooting. But a cheap meter is fragile. In other words, be ready to buy another one after you drop or fry this one.

[Interesting but non-essential information. Not all that long ago, electronic meters simply didn't exist. The only meters we had were *analog*, where you get your readings on a scale like an automobile

speedometer. A needle points to the right numbers. But the analog meter has several scales of numbers. That can make it hard to read. And it's fragile. If you hook it up wrong, it makes a pathetic little squeak and dies—reminds me of the sound the frogs made in biology class when we killed them.]

How to Use a Meter

Sometimes we need to use a meter to find out what's going on in a circuit. I know there are plenty of old timers who size up a circuit by the sensation in their hands or fingers. But go ahead, be a wimp—use a meter. If nothing else it'll save you a lot of time.

If you're new to using a meter, or if you're using a meter you don't know well, begin by measuring things you already know—like a 9 volt battery, or a 120-volt house wall socket.

A meter is called a *multi*meter because it has multiple functions. With one meter you can usually measure volts (V), amps (A), and ohms (Ω). You can measure each of these three for alternating current (AC) or direct current (DC). That gives six possible functions. It's up to you to select the function you need.

> [Interesting but non-essential information. Actually there are only five functions because the continuity/resistance function is the same for AC and DC.
> That's because the circuit's power is removed and the meter's battery supplies the power for checking resistance/continuity.]

Troubleshooting with a Meter

	Voltage (V)	Amperage (A)	Continuity/ Resistance (Ω)
AC	x	x	x
DC	x	x	x

Figure 7. Multi-meter functions

In control circuits we use primarily AC, so let's limit our selections to alternating current functions. These functions will be marked either AC or ~ (the AC squiggle).

Voltage

Voltage is indicated by the letter V. The alternating current voltage function is marked either by *VAC* or the letter V with a squiggle above it. We call the voltage part of the meter a *volt*meter. Set your meter for the highest possible voltage setting. You may think you know what voltage to expect, but you could be wrong. After the meter tells you what the voltage is, you can move down to the range that the voltage falls into.

On your meter you'll find two *leads* or *probes*. One is black and one is red. For our purposes here when measuring AC, it doesn't matter which you put where. It matters only for direct current (DC) and applications where polarity it critical. Make absolutely certain that your fingers touch only the colored, insulated part of the probes. You can get shocked if you let your fingers stray onto the metal tips.

To check for line voltage coming into the circuit, put one lead on L1 (black) and one on L2 (white).

QUICK & BASIC TROUBLESHOOTING

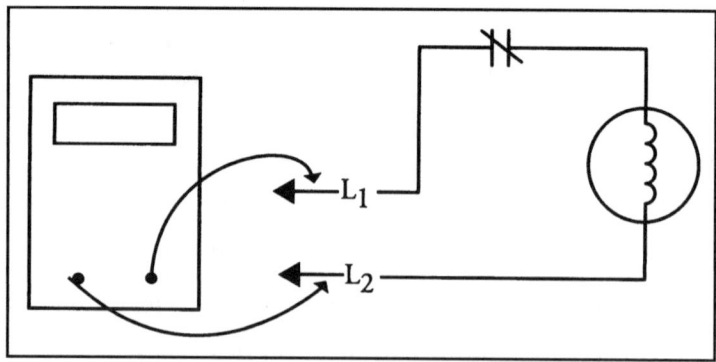

Figure 8. Using a voltmeter to check for voltage

If you're measuring voltage coming into a control circuit transformer, you may very likely find a meter reading within a few volts of 120V or 240V. Both of those are standards for "line" voltage. It doesn't have to be exactly one of these numbers, but close. In fact the purpose of this step is often to find out if there is any voltage at all. If there is, you can say, "Yep, we got power."

[Interesting but non-essential information: the voltage delivered by the utility can vary depending on time of day or time of year, or location. When there is a greater overall demand (e.g. during high air conditioning demand), your voltage may be lower within 10%. As the voltage going into your transformer primary side goes up or down, so does the voltage coming from your transformer secondary.]

Take a moment to notice in the drawing below that positioning the meter to measure voltage puts it in a parallel circuit with the load.

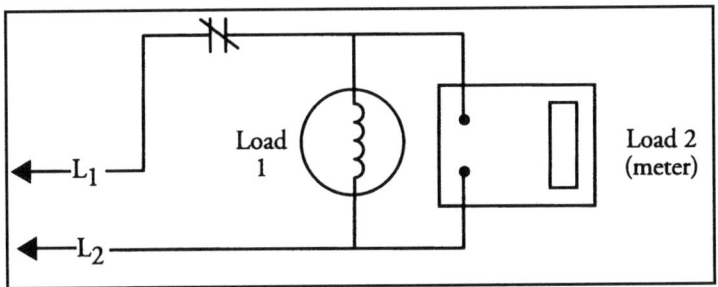

Figure 9. Voltmeter in parallel with the load

Continuity (Resistance)

Perhaps you've watched your buddies take a meter from the case, touch the two probes together, and make the meter beep. The beep is the meter's way of saying, "I found a path to send out electricity and get it back." This is called *continuity*. To do a *continuity check,* set the meter on the omega (Ω) setting. Some upper-end meters have a separate continuity setting, but it's not necessary.

> [Interesting but non-essential information. The omega symbol is used because the letter "O" (for *ohms*) looks too much like a zero. So we use the Greek letter for "O," which is the omega.]

Remove all power from the circuit before taking any measurements with the ohmmeter or continuity functions. Use the voltmeter to confirm that there is no voltage. This is important because the battery in the meter is the power supply when you set the meter for ohms or continuity. You don't get to select AC or DC because the meter is supplying the power. See the diagram below for the power supply-switch-load configuration when using the ohmmeter. With some inexpensive meters, if you try to check continuity with high voltage, you may kill the meter.

When you touch the two meter probes together for a continuity check, the path of the electricity from the meter's battery is simply from one probe to the other—big deal. But imagine how useful it could be to find out if there is an unbroken path through a switch or load. You touch one probe to each side of the switch or load. If the meter beeps, it has found a path. That usually means that the switch is closed, or that the load is good.

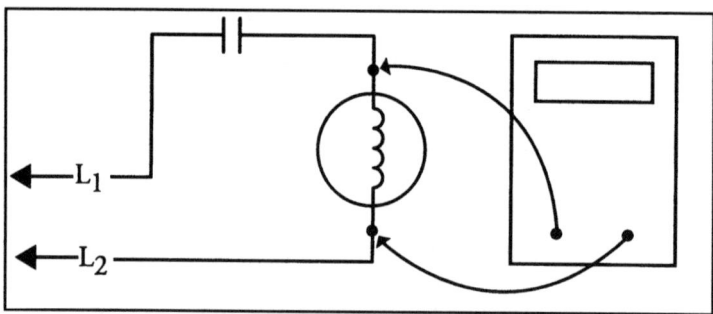

Figure 10. Using the ohmmeter—first remove power

We might also use this continuity check to see if an insulated wire is continuous or if it is broken inside.

> [Interesting but non-essential information. Multi-stranded copper wire may show continuity if only one tiny strand is unbroken. But the load won't operate.]

When we did these continuity checks, the meter was actually looking for resistance to the flow of electricity. The meter gave a numerical display in ohms. In controls circuits we usually aren't concerned about the actual number. Any amount of resistance on the display is the equivalent of a beep. Some meters don't beep, so the presence of any ohm measurement says there's continuity.

Amperage

Amperage is another meter function that we use in troubleshooting. It uses the ammeter portion of the meter. An ammeter measures amperes, or amps. Amps are abbreviated as "A".

Here's a caution. Our control circuits are AC amperage. Cheaper meters may have a function for only DC amperage. If you use the DC setting on alternating current, your meter may squeak and die.

We use the ammeter to determine the circuit's amp draw after the circuit's been running a few minutes. The main thing we use this measurement for is to set the thermostat anticipator for ideal room comfort. (More on this is coming up in the section about troubleshooting switches.)

Every load has an amp draw, or *amp rating*. You can find the amp rating printed on a load itself, on its box, and in the product data sheet. Finding the amp draw of a circuit can be as simple as adding the amp ratings of all the parallel loads.

To isolate a problem load, you can allow only one load to come on at a time to see which load is overdrawing the circuit, for example, the burner, the fan, or the humidifier.

In controls work you can expect the amp draw to be in tenths of an amp. It's easiest to use a digital meter that measures in decimals. And the lock feature is a blessing. You can lock the reading from a hard-to-see place and bring it back to where you can read it.

There are two ways to measure amps with a meter. The easiest is to use a meter with a clamp either built-on, or as an attachment. Place the clamp around any wire in the circuit to get the amp reading. Make sure it's just one wire.

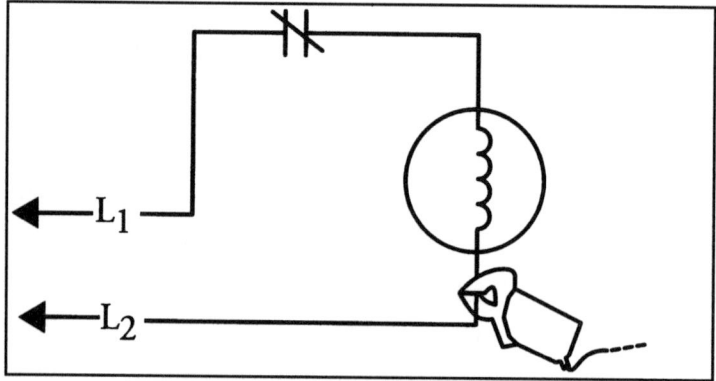

Figure 11. Clamp-on ammeter

If your meter doesn't have a clamp, you must place it in series with the circuit. That means you must break the circuit and put the ammeter in it.

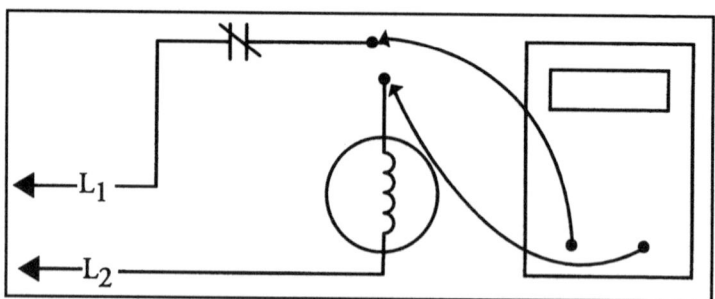

Figure 12. Using an ammeter in series

Troubleshooting with a Meter

Sometimes the easy way to get the ammeter in series is to place the probes on the terminals of a switch, such as the thermostat. The ammeter there will have the effect of closing the switch.

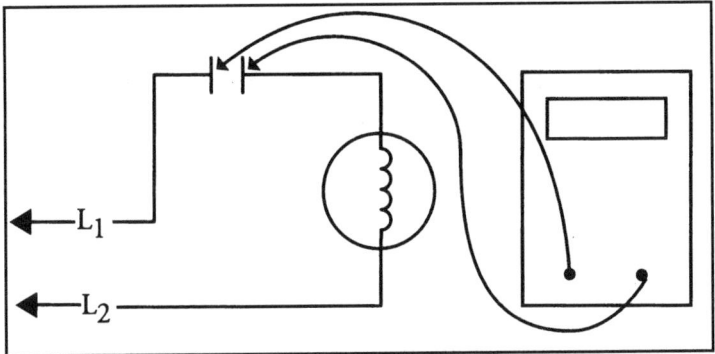

Figure 13. Ammeter in series across a switch

If your meter won't measure tenths of an amp, here's what you can do. Make a coil of ten turns of a piece of normal thermostat wire. Put the clamp portion of the meter through the coil. Put the ends of the coil in series with the circuit. Divide the meter reading by ten to get your final circuit amp draw.

Troubleshooting Individual Components

Let's look at troubleshooting individual components—the power supply, switches, and loads. After we get that down, we'll move on to troubleshooting these components in circuits.

Troubleshooting the Power Supply

When we troubleshoot a power supply, the first thing we need to find out if there's voltage and how much. The power supply in a control circuit is the transformer. The purpose of the transformer is to reduce line voltage to low voltage for use in a control circuit. Line voltage is the same electricity found in ordinary wall sockets. Low voltage is a fraction of line voltage. For example 24V is one-fifth of 120V.

The transformer *primary* is the side where line voltage goes into the transformer. The *secondary* is the side where the electricity comes out as 24V for our control circuits.

Troubleshooting Individual Components

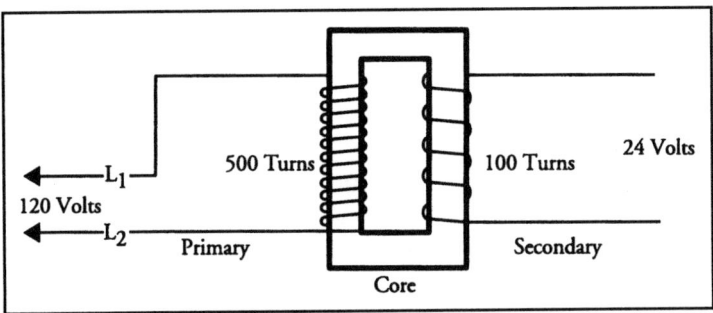

Figure 14. Step-down transformer

To confirm that the power supply is good, set the meter for AC voltage, VAC, or V with a squiggle over it. Put one meter probe on each of the two terminals of the transformer secondary side. Those two terminals are usually labeled R and C. If you find 24V or slightly more, you know the transformer is good.

If you find no voltage on the secondary side, or less than 24V, measure the voltage on the primary side. If you find line voltage on the primary, and no voltage on the secondary, the transformer is bad. That's because voltage going in should come out the other side. If you have 120V in and less than 22V out, it's worth replacing the transformer. But if you find less than 120V on the primary, then less than 24V on the secondary makes sense because the secondary is proportional to the primary—less in means less out.

No voltage into the primary means you don't have incoming power. Check the circuit breaker or fuse at the building's incoming power. You can also use a meter to check continuity of the primary and secondary sides of the transformer. No continuity confirms a dead transformer.

> [Interesting but non-essential information. Many pieces of modern equipment require correct polarity. When polarity is reversed, that is, if L1 and L2 are reversed, the equipment may light and then lock out. White neutral on the equipment should be zero volts to ground. The black L1 should be 120V to ground.]

Types of Ground

Earth ground means that the electricity can follow a path and eventually get back to the earth. *Chassis ground* or *equipment ground* is earth ground brought to the furnace or boiler by conduit or a separate ground wire.

> [Interesting but non-essential information. Sometimes the terms *ground* and *neutral* get interchanged. Ground is either to earth ground, or to the water meter. Neutral is the electrical neutral at the fuse box.]

Another type of ground is *burner ground*. This type is usually what's meant when we're dealing with electronic ignition and burner circuits. In this case, there needs to be a wire that connects directly from the ignition to the burner. That's so that the ignition signal can get all the way through the circuit to confirm that flame is present.

> [Interesting but non-essential information. Some older equipment used the furnace sheet metal as a return

path. This sometimes gets corroded. Add a separate burner ground wire to fix this problem.]

Why do Transformers Go Bad?

There are several common reasons why transformers go bad. All are within our control except for lightning. A cause within human control is that somebody tried to put too high a voltage of electricity into the primary side. An example is applying 240V to a transformer rated for 120V. Another cause is that someone tried to get too much out of the secondary by connecting too large a load or too many loads to it. A third reason is that someone "shorted out the load," so that the transformer became the load and "burned."

What goes bad in the transformer is a *winding*. A winding is a spool of very thin metallic thread. Too much electricity kills the winding. In the industry we call that "smokin' the transformer" or "lettin' the smoke out." No matter how hard you try, you'll never convince your supplier that it came out of the box that way!

"Short out the load" means that somehow the load's (for example gas valve or zone valve) two terminals got connected together. That can easily happen if you let bare wire tips touch each other when the circuit is powered.

The other instance of burning out the secondary is when the load or loads are demanding more electricity than the transformer can deliver. A transformer's capacity is given as a *VA rating*. VA means volts multiplied by amps. A 40VA transformer is typical. This means that the circuit's total amps multiplied by 24V (a given) must not exceed 40 VA. If the loads in the circuit exceed the transformer's VA rating, the transformer dies.

To figure the VA of the circuit, add the amp draw (can be found on the load itself) of all of the loads, and multiply by 24V. The VA rating of the transformer has to be greater than the combined VA of the loads.

Here's a typical situation where the transformer burns out. We have a 40VA transformer. It's been fine for years. In the middle of the coldest darkest night in the coldest cold spell ever, the transformer fails. Perhaps you can see and smell that the secondary has burned. Your voltmeter may tell you that you have voltage at the transformer primary, but none coming from the secondary.

It's tempting to replace the transformer and be done with it. But there's a reason for the failed transformer. The new one may burn out, too. Let's say that there are six zone valves connected to that transformer. Each valve has an amp rating of .32A. When we multiply .32A times six valves, we get 1.92A. We multiply 1.92A by 24V and get 46VA—too much for a 40 VA transformer.

Why didn't the transformer fail sooner? Because it took the coldest darkest night for all the zone valves to call for heat at once.

Let's look at the VA rating another way. If we have a 40VA transformer, how many of those .32A zone valves could we put on it? Multiply .32A times 24V. That gives us 7.68VA. Dividing 40VA by 7.68VA gives 5.2 zone valves.

Because we must not exceed the transformer's VA rating, we would use 5 or fewer zone valves.

To get more VA, an installer will sometimes wire two transformers together. But it is highly preferable to leave them separate for two reasons. First, if one fails, you still have the other one working. Second, if they're wired together, you have to be sure that they're *in phase*. That means the *hot* side of the first transformer is wired to the *hot* side of the second transformer. And of course the *common* side of the first transformer is connected to the common side of the second transformer.

Troubleshooting Switches

When we troubleshoot a switch, we're trying to find out if it's open (off) or closed (on).

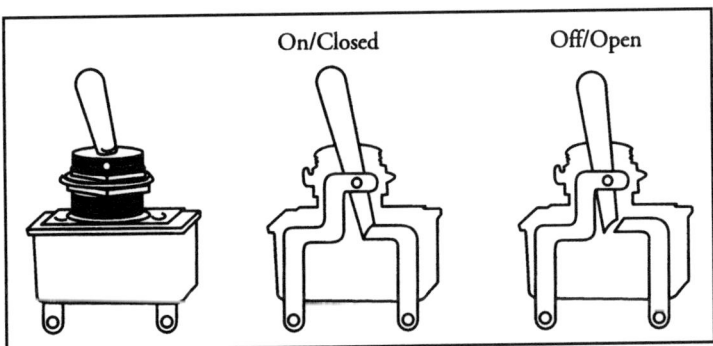

Figure 15. Simple switch, closed and open

[Interesting but non-essential information. *Open switch* and *closed switch* may not mean what you think. You

QUICK & BASIC TROUBLESHOOTING

may be used to thinking about valves and piping—open means stuff flows, and closed means it doesn't. In electricity, think of the switch as a drawbridge instead of a valve. An open drawbridge interrupts the flow of traffic. A closed drawbridge allows traffic to move. So in electricity, a closed switch or a closed circuit is what we need for the circuit to work.]

A continuity test is good to find out if a switch is closed. A closed switch is a continuous path. If the switch is closed, the meter will give a beep or an ohm reading.

A second way to test a switch is using the voltmeter. In this case keep power to the circuit. Measure across the two switch terminals. If 24V appears on the meter display, it means the switch is open. If there is voltage in the circuit, but no voltage reading across the switch, the switch is closed.

[Interesting but non-essential information. Here's why no voltage reading means the switch is closed. A voltmeter is actually measuring *voltage drop*. Voltage drop is the difference between the amount of voltage at two different points.

If the switch is closed, there's a continuous path. A switch lets electricity pass almost effortlessly. Since electricity is passing effortlessly, there's no difference between the amount of electricity before and after the switch. There's no drop in the amount of electricity.

If the switch is open, there is a voltage drop. Think of the open switch as a drawbridge. Electricity piles up behind the open switch. The voltage drop is the difference between the amount piled up on one side (24V) and none on the other side.]

Troubleshooting Individual Components

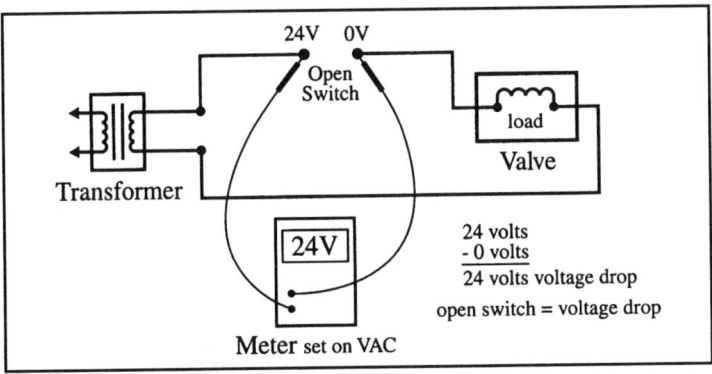

Figure 16. Voltage drop with a switch

A third way to test a switch is to "jumper it out." When a switch seems to be open when it shouldn't be, simply connect the two switch terminals together with an alligator clip jumper wire. If the circuit then works, you have confirmed that the switch was open. The jumper gave the electricity an alternative path around the switch.

Here's an example. When you suspect that you have an open limit switch that should be closed, you can put a jumper across the limit switch terminals. If the circuit then works, it means you have a bad limit. Remove the jumper. Replace the limit, and your job may be done.

But, there's usually a safety reason behind an open limit switch. Limits seldom go bad from age. It's wise to find out why a limit failed before you leave the job.

An open switch is just like an opening in a wire. A closed switch is just like a continuous wire. When you measure voltage across a switch, if there is voltage, the switch is open. If there's voltage to the switch, but no voltage reading across it, the switch is closed.

Troubleshooting Loads

A load is a consumer of electricity. It uses electricity to create heat, motion, light, sound, etc. The load uses electricity by offering resistance. That happens because the inside of a load is usually a coil. A coil is much harder to pass through than a straight path. If the coil becomes broken, or open, electricity can't pass through. Since we can't see inside, we use a meter to test the load for resistance.

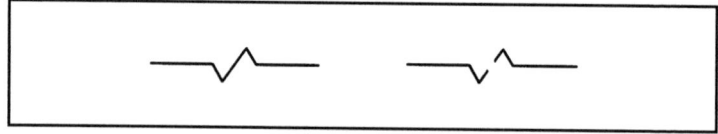

Figure 17. A "good" load, and an "open" load

To test for an "open" in the load, select the continuity (ohms) meter function. (See the ohmmeter/continuity portion of the meter section of this book for details.) Remove power from the circuit. Put the meter probes on the two wires or terminals to the load. If there is a beep or ohm reading, there is a continuous path through the load. The load is good.

> [Interesting but non-essential information. A voltage test doesn't work on an individual load. That's because the meter will show a voltage drop for both a good load—because it consumes electricity, and a bad load—because it has break or opening.]

Troubleshooting Relays

Part of a relay is a coil, which is a load. A relay is also made up of switches.

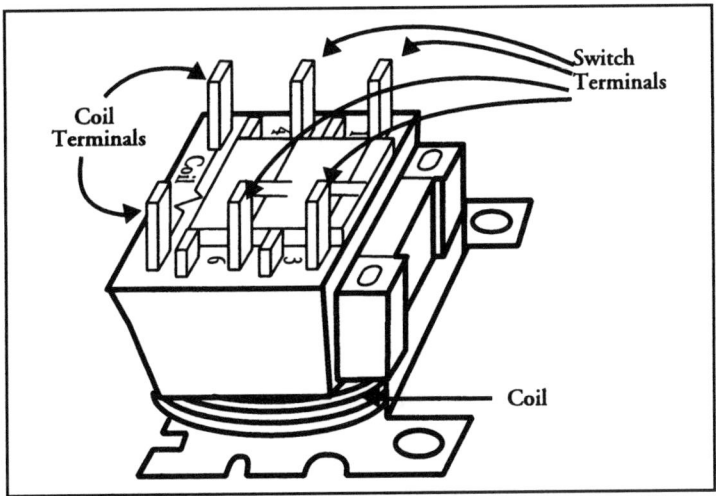

Figure 18. A relay

Even though the coil is at the bottom of the relay, the terminals for the coil are often on the top. You can visually follow the coil wires to their terminals. You can use these two terminals to perform a continuity test of the coil. Remove power first. If the coil is good, you'll hear the meter beep.

Relay switches appear as pairs of terminals. Pressed into the plastic of the relay are symbols that show each switch as "normally open" (NO) or "normally closed" (NC). "Normally" means when the coil is not powered. If you're not comfortable with relays yet, this could be one reason why. As soon as there's power to the relay coil, the switch positions are the *opposite* of the designations on the relay.

QUICK & BASIC TROUBLESHOOTING

A switch marked normally open should be closed if the circuit is powered.

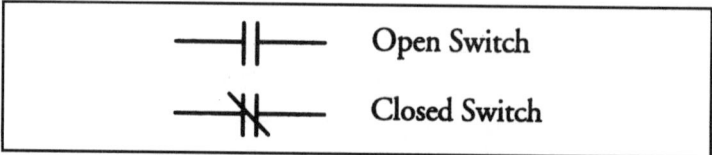

Figure 19. Switch symbols

Notice whether the switch you're testing is marked normally open or normally closed. Use the continuity function of your meter to test if the switch really is open or closed. You had to remove power before you did the continuity test, so you can expect the switches to be in the position they are marked.

Now power the relay coil. Change your meter to the voltmeter function. Measure across that same switch. A voltage reading (24V or 120V, depending upon the circuit the switch is in) means the switch is open. No voltage means the switch is closed. Since the relay coil is powered, the switches should be in the position opposite to that marked on the relay.

Troubleshooting Panels

Panels have a reputation for making things difficult. The reality is that they sort things out and make life easier. Like in the story at the beginning of this book, panels often are unfairly blamed for troubleshooting problems.

Troubleshooting Individual Components

When there's no troubleshooting chart, often the best troubleshooting method is to remove everything except power to the panel. (Label every wire before removing it!) Then see if the panel operates normally with everything removed. If yes, then there's probably nothing wrong with the panel itself. The source of the problem is probably in one of the components attached to it. Or the problem may be a wiring error. Or the problem may be a short in the wiring.

With everything but power removed from the panel, add one component, say a thermostat. If the panel works, add a load. Keep adding back components systematically until the panel fails. Chances are the last component you added before the failure is the *key* to the failure. The problem could be in the component itself, a short in the wiring to the component, or an error in how the component is wired into the panel.

Check your wiring. Do the wires from the component really go where they're supposed to on the panel? Make sure that no bare ends of wires are straying to touch anything but the terminal they're to be connected to. Look for damage to the insulation, or a nail or staple into the wire.

Troubleshooting the Complete Circuit

Troubleshooting using the "Home Run," "Hop-Scotch," "Leap Frog," or "Daisy Chain" Methods

The idea behind this troubleshooting method is to find where electricity should be and isn't, or shouldn't be and is. That's where the problem will be. It could be a broken wire, an open switch, or a break in the coil of the load. A systematic approach is preferable to random testing of components, or—heaven forbid—changing out parts

Let's say we have a circuit of transformer (power supply), thermostat (switch) and valve (load).

We know there's something wrong in the circuit because the valve won't open.

Before you do anything else, turn up the thermostat as high is it will go. This is to make certain that the switch is closed. (Closed thermostat = call for heat.)

Troubleshooting the Complete Circuit

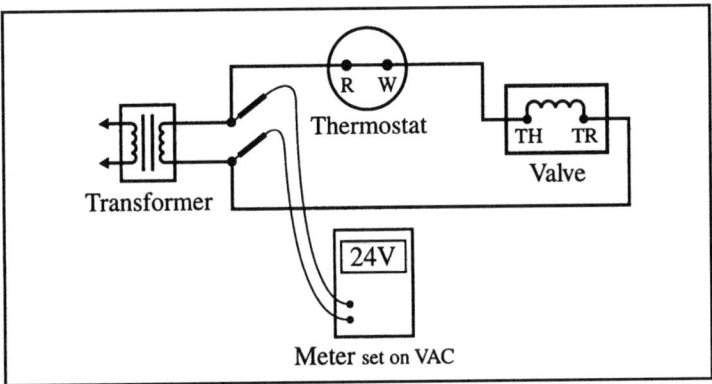

Figure 20. Using the "hop-scotch," "home run," "leap frog," or "daisy chain" method to test the transformer in a circuit

We've already attached the first meter probe to one terminal of the transformer secondary. With the second probe, touch the second transformer terminal. A reading of 24V confirms that the transformer is good, and that we have power to the circuit.

If there weren't 24V on the secondary side of the transformer, we would have tested the primary side of the transformer to see if there was power (120V since the primary side of the transformer is line voltage). If there were, it would mean we had a bad transformer (power available to go in, but none coming out). If there wasn't power going into the primary, then the problem is with power into the circuit—time to check the circuit breaker or fuse to the circuit.

Here's the most important part—keep the first meter probe clipped to the low side (common) of the transformer.

QUICK & BASIC TROUBLESHOOTING

Place the second meter probe on the closest side of the next component in the circuit, the thermostat. (In a real life situation, you very likely can't get to the thermostat itself [probably in the living space] from where you and the transformer are [in the basement, crawlspace, equipment, closest and other uncomfortable places that HVAC people hang out].) We'll take care of this obstacle in a moment.

Touch the second meter probe to the first side of the thermostat. That's going to be the terminal labeled "R."

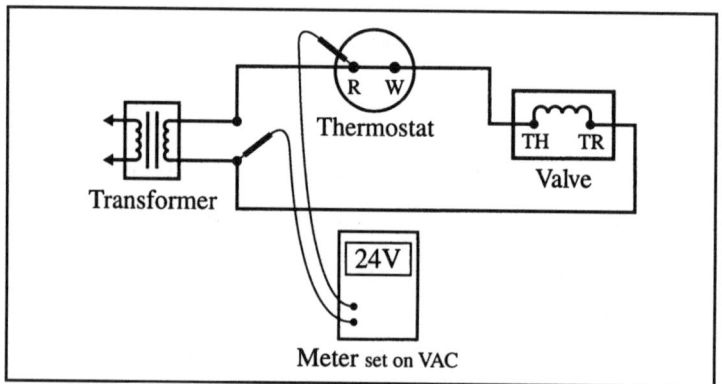

Figure 21. Testing wiring from the transformer to the thermostat

[Interesting but non-essential information. There are two thermostat terminals for heating—R and W. Typically, R is connected to power, and W is con-  nected to the load. R stands for red; W stands for white. Granted the wires aren't always these colors, but they usually are. If it helps you remember, you can think of W as also standing for "warm," which is usually what we want the load (valve) to deliver.]

Troubleshooting the Complete Circuit

[More interesting but non-essential information. There are still many three-wire controls, especially in the oil industry. An example is the RA117 stack relay. Fortunately, the thermostat has usually been switched to a two-wire 24-volt style. It's connected to the B and W terminals of the stack relay—the W is left out. One way to remember which terminals to use is W-B means "will begin."]

We're testing the portion of the circuit between the first transformer terminal and the first thermostat terminal. Since we already confirmed that the transformer is good, what we're interested in right now is the wire between the transformer and thermostat. If 24V shows on the meter, it means the wire and the connections are good.

Now move the second meter probe to the second side, or second terminal (W), of the thermostat. If you see 24V on the meter, it means the switch is closed, and the thermostat is good—electricity has gotten through the thermostat.

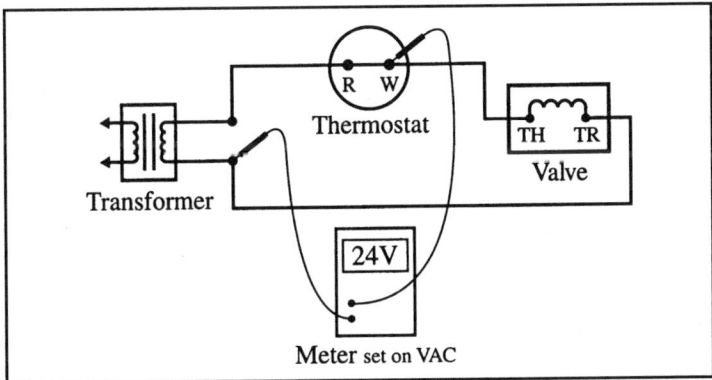

Figure 22. Testing the thermostat

QUICK & BASIC TROUBLESHOOTING

But if you don't find 24V at this point, there are a couple things to do. First, make absolutely certain that you've turned the thermostat up as high as it will go.

If you're certain you have a call for heat and you don't have 24V here, you may have a bad thermostat. Let's pause to test the thermostat alone. Take the first meter probe off the transformer and put it on the R terminal of the thermostat. If the thermostat has a subbase, consider it as part of the thermostat. Keep everything else wired up—there is still power to the thermostat. Put the second probe on W. This time there should *not* be a voltage reading between R and W. If there is, it means the thermostat switch is open. And if you're certain you've set it for a call for heat, this means the thermostat is bad, because the switch in it won't close. But don't replace the thermostat yet. (But whenever you replace a thermostat, save yourself lots of trouble and replace the subbase as well.)

Think or go back to the section in this book about testing switches. You're testing just the switch (between terminals R and W) and you have voltage across it. Since it's a switch, it doesn't use power—it just passes it through, like a drawbridge. That means there will be no voltage drop between R and W. Voltage drop is actually what is being measured when we see a voltage reading on the meter. Again voltage drop is the difference between the amount of electricity going in and the amount going out. A switch doesn't use any electricity. It just lets it pass through. So there's theoretically the same amount leaving the switch as there

Troubleshooting the Complete Circuit

was coming in to the switch. (The thermostat's anticipator is a load, and it uses a small amount of electricity to create a small amount of heat.)

[Interesting but non-essential information. Power-stealing thermostats show voltage to W, Y and G with no call for heat. The amps are so low that nothing happens, but you can read 24 volts. In this case a test light is more useful than a meter because there aren't enough amps to light the light.]

If a switch is open, no electricity passes through. There is electricity waiting to pass through, but if the switch (drawbridge) is open, the electricity is just sort of piled up on one side of the bridge. There is no electricity on the other side of the open drawbridge, because there's no way for it to get there if the switch/drawbridge is open. So, again, when we measure voltage, we're really measuring voltage drop. Voltage drop is the difference between the amount of electricity on one side versus the other side. A switch, whether open or closed, never uses electricity. But an open switch does show a voltage drop because there's electricity piled up on one side, and there's none on the other side.

Of course you could choose instead to test the switch using the ohmmeter portion of the multimeter. In that case you would remove one of the wires to the thermostat in order to remove power. Set your meter to ohms (Ω). Touch one meter probe to the thermostat R terminal and the other

QUICK & BASIC TROUBLESHOOTING

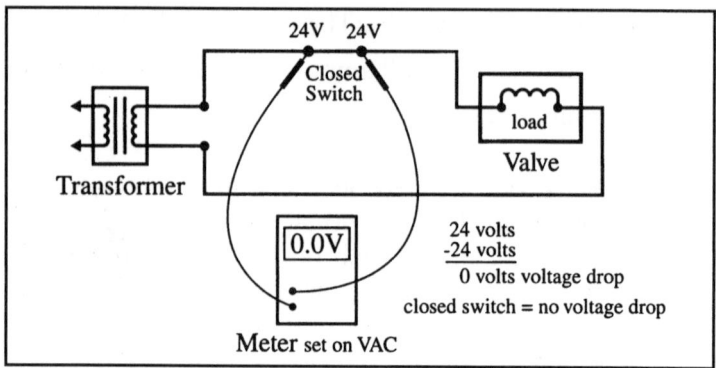

Figure 23. The voltage drop drawbridge

to W. If the switch is closed, the meter will beep or show an ohm reading. That means the meter used its battery to send power through the switch. The meter sensed the power come back. That confirms that there is a path for the electricity to go all the way through the switch. If the meter doesn't beep, the switch (thermostat) is open. If a thermostat is set to call for heat (turned all the way up), and the subbase switch is set for "heat," and the thermostat tests as open with the meter, it is probably a bad thermostat.

Here's another thing you can do to. Think back to where we had one meter probe on the first side of the transformer, and the second probe on the W terminal of the thermostat. If you don't get 24V at this point, you can choose to simply *jumper* out the thermostat.

Caution! A switch is the only component of a circuit that you can safely jumper out. If you jumper out a load or

Troubleshooting the Complete Circuit

power supply, you're likely to burn things up. For that reason it's better to use a 3-amp circuit breaker instead of a jumper wire. Then a mistake just trips the breaker.

Jumpering out the switch means you take an insulated wire with an alligator clip on each end, and attach one clip on R and the other on W. The effect is that the jumper is an alternative path for the electricity to take around the switch/drawbridge. Now, with the jumper attached, see if there's 24V from the first side of the transformer to thermostat terminal W. If there is 24V, it means the thermostat is open. If you're certain you have the thermostat turned all the way up (call for heat), you have a bad thermostat.

Here's why you're inviting trouble if you jumper out a load or a power supply. Electricity needs to be consumed, or resisted, by something in the circuit. That's what the load does. The load slows it down to the point where all its power has been spent. It won't damage the rest of the circuit as along as there's a load to use up its force. If you jumper out a switch, electrically the effect is nothing other than closing the switch. A closed switch is just a short piece of wire. But if you jumper out the load (and if there's no other load in the circuit), there's nothing in the circuit to use the electricity.

To electricity, any coil looks like a load. When you jumper out the load coil, the electricity will look for any other elements in the circuit that it can use as a load. These will burn if they get all of the electricity in the circuit. There's a winding of very small wire in the thermostat,

QUICK & BASIC TROUBLESHOOTING

called an *anticipator*. It will burn when the intended load in the thermostat circuit is shorted

The anticipator is intended to get a tiny amount of electricity. It uses that electricity to create a tiny amount of heat to trick the thermostat into thinking the room is warmer than it actually is. The effect is keeping tighter temperature control of the room. But without the larger load, all the electricity is too much for the anticipator to handle. The anticipator burns with a bit of smoke and a distinctive burned electrical smell. The thermo-stat no longer works. Even if you don't know how it happened, thermostats just don't come out of the box that way.

The transformer is made up of two coils. The secondary coil is the power supply for the control circuit. If you short out the real load, the electricity may find the transformer coil to be an attractive substitute.

Whether you short out just the intended load, or somehow connect the two transformer terminals together, you still have electricity looking for a load. If the loads connected to the secondary side of the transformer are shorted, a transformer coil burns. Again you have that distinctive burned electrical smell.

A *fuse* or *circuit breaker* is a very practical and inexpensive way to protect your circuit. Transformers are available with a built-in circuit breaker. They cost just a few dollars more than a good quality standard transformer. That means that instead of replacing the transformer, you just reset the breaker. Another choice is to insert your own fuse (talk to your neighborhood

Troubleshooting the Complete Circuit

electronics store) or fusible link into the circuit, especially when you're troubleshooting. Fuses are much cheaper and easier to replace than transformers and thermostats.

Moving on, let's say that the thermostat tested good. Make sure that the first meter probe is still on one transformer terminal. Put the second meter probe on the closest valve terminal. On a gas valve, this terminal may be TH. This means it should be connected to the thermostat. On a zone valve, it would be either of the two motor leads or terminals.

In this case we're testing only the wire between the thermostat and the load. There should be 24V on the meter.

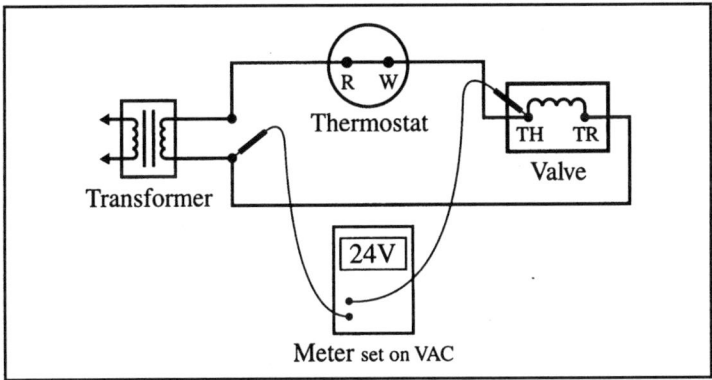

Figure 24. Testing the connection between the thermostat and the valve

Move the second probe on to the TR terminal, or the connection to the transformer. There should be 0V because the meter sees no voltage drop from TR back to the transformer.

To test the value, remove power, set the meter for ohms, and, perform a continuity check.

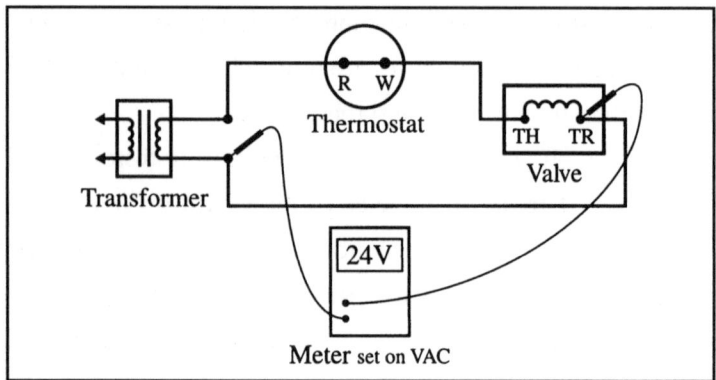

Figure 25. Attempting to test the valve

Here's a note about motor windings. Sometimes after they fail, they can work again—usually for just a little while. This is called an *intermittent,* and we hate it.

[Interesting but non-essential information. There are devices available that record what's happening when you're not there.]

Here's how an intermittent motor might behave. Due to overheating of some sort, the motor winding expands and breaks. It becomes "open." The motor doesn't work. When the motor cools, though, the ends of the break may touch and reconnect themselves—temporarily. The motor will work again until it heats enough for the broken ends to expand apart again. The solution is to replace the motor.

We have tested our way all around the circuit. One of these tests should have identified where the problem is.

Troubleshooting the Complete Circuit

A very unexotic requirement of troubleshooting sometimes is simply checking your wiring. The questions are, "Does it really go where I think it does," and "Is where I think the wiring should go really where it should go?" Fortunately, you can probably check both of these at the same time.

This is a hands-on process. Here's what you do. Let's say you have a two-zone zone valve system.

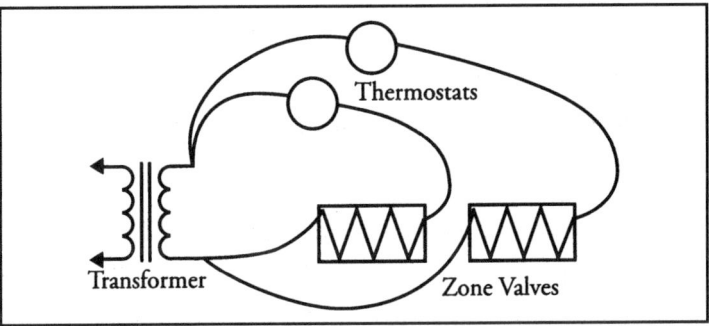

Figure 26. A two-zone valve system

Think about a complete circuit: power supply, switch, load, and back to power supply. Don't let the spaghetti of the actual wiring distract you. You have to look at only one thing at a time. Here it is.

- Where's the power supply?
- Where's the switch (thermostat) for this circuit?
- Which wire connects this power supply to this switch (thermostat)? Gently tug the transformer end of the wire to make sure it's firmly connected. Trace it with

QUICK & BASIC TROUBLESHOOTING

your fingers to make sure that it actually connects to the thermostat R terminal. Gently tug the wire at the R terminal to make sure it's firmly connected. Look out for loose screws and wire connectors.

Note: there may be more that one wire coming from one side of the transformer. If there are two, only one of them will connect to the thermostat in question. There may also be wires connected by a wire nut. In the case of a wire nut, there's a wire going from the transformer to the wire nut. And there's a wire going from the wire nut to each of the thermostats. Don't let the number of wires confuse you. There are three wires to connect two thermostats to the transformer—one wire to connect each of the three components to the wire nut.

- There's another wire connected to the W terminal of the thermostat. Is it firmly connected? Trace that wire with your fingers and see where it goes.
- Let's say the wire goes to the zone valve. Is it firmly connected to the zone valve motor? (Disregard any wires that go to a zone valve end switch, that is, not to the motor.)
- There should be a second wire coming out of the zone valve motor. Trace that wire back to the transformer. (It's a different transformer terminal that you started out with, but it's still on the "secondary" side of the transformer.)

This is your complete circuit. You just completed two checks. You made certain that you have a power supply,

Troubleshooting the Complete Circuit

switch, and load. And you made sure that there are wires firmly connecting all the components in a circuit (circle that comes back to where it started).

Since we're talking about a two-zone circuit, we have another set of power supply, switch, and load to trace. In this system we're using one transformer. So we start with the *same transformer*.

[Interesting but non-essential information. Because these two circuits share a power supply, they are technically one *parallel* circuit. But we can think about them as two separate circuits that happen to share a power supply. Like two kids with the same mom, they're separate, but they get fed at the same table.]

On the first side of the transformer, find a second wire. Make sure it's firmly connected to the transformer. Trace the wire with your fingers to make sure it actually goes to the second thermostat R terminal. From the second thermostat W terminal, trace the wire to the valve. From the valve trace a new wire back to the other terminal on the transformer secondary. Arriving back to the power supply assures that you have a complete circuit, with the three required components of power supply, switch, and load.

Here's a different way of tracing the wires of a parallel circuit. Make sure every 24V load is connected directly to the "C" common side of the transformer. (All 120V loads connect to the neutral—L2.) Make sure all the switches connect to the "R" side of the transformer.

QUICK & BASIC TROUBLESHOOTING

Troubleshooting Using a Troubleshooting Chart

A troubleshooting chart is a flow chart. It shows a sequence of steps, in a particular order. *You may not skip around. You may not start in the middle. You may not skip steps.*

Here's a fun non-technical troubleshooting chart. It certainly isn't an example of the best way to be, but it gets a chuckle out of most of us. Let's use it to get used to how a troubleshooting chart works.

The pattern is

- A question
- A choice of yes or no
- An action to take or a new question

Always start at the top.

The first question on this silly chart is—"Does the thing work?"

You choose "yes" or "no."

Choosing "yes" takes you to the action "don't mess with it," and on down to the final resolution "No problem."

Choosing "no" takes you to a new question—"Did you mess with it?"

The answer "yes" takes you to a piece of information ("You fool") and to a new question—"Does anyone know?"

Again there are two possible answers to the question. And those lead in two possible directions. The answer "no"

Troubleshooting the Complete Circuit

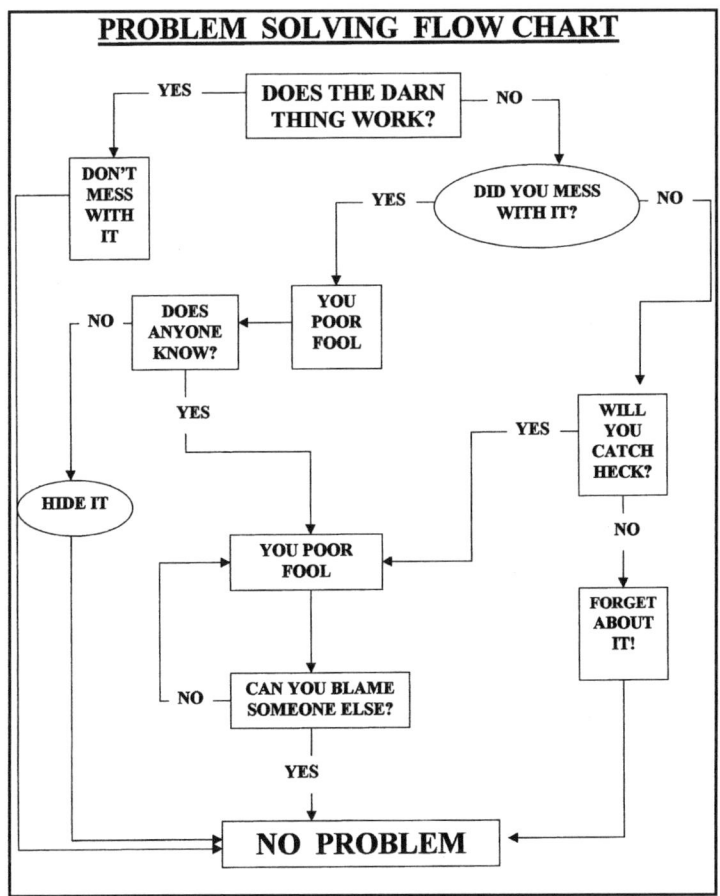

Figure 27. A silly troubleshooting chart

leads to the instruction "Hide it" and to the final resolution of "No problem."

Choosing "yes" for the question "Does anyone know? Leads down an entirely different path. There's the piece of information "You poor fool" and a new question "Can you blame someone else?"

QUICK & BASIC TROUBLESHOOTING

If you can answer "yes," you move on to the resolution "no problem."

The answer "no" leads back to "you poor fool" and back to "can you blame someone else?" and so on in a pattern called an *infinite loop*. You just keep going around and around. A real troubleshooting chart shouldn't have any infinite loops in it. But seeing it in this one feels a little like real life, doesn't it? The lesson is that you can't get out of the loop until you answer the question "yes." Then you can move on to "No problem."

Let's go back up to the top of the chart and take the remaining path—the "no" answer to "Did you mess with it?"

Following the "no" answer is the question "Will you get in trouble?"

A "yes" answer to "Will you get in trouble?" takes us to the piece of information "You poor fool." We've already seen the rest of that path.

Moving back up to the "no" answer to "Will you get in big trouble?" there's only one piece of information "Forget about it!" And that moves directly on to "No problem."

The method of using a technical troubleshooting chart is the same. You have to start at the beginning. The answers you give to the questions determine where you move to next in the chart. Eventually the chart leads you to the resolution of the problem.

Especially if a control is electronic, the problem often isn't what our "logical" thinking suggests it might be. A

Troubleshooting the Complete Circuit

classic appears in electronic ignition—the gas valve won't open. Our logical thinking, without the help of a chart—says that the gas valve won't open because: 1) it doesn't have power, 2) the switch doesn't work, or 3) the valve is bad. So we use a meter to confirm that there is indeed 24V coming to the gas valve. Then we conclude that the valve must be bad because it has what it needs to open, and it isn't opening. The problem is that we're wrong. All that changing the gas valve will do is keep us busy. It won't fix the problem.

[Interesting but non-essential information. Manufacturers say that of the product they take back on warranty, about 80% has nothing wrong with it. They take it back to earn your good will. What they can't do for you, though, is replace the time (translate "money") you spent replacing something that wasn't defective.]

Let's take a moment and consider why changing parts sometimes works. Many of the problem jobs, especially on new installations, are faulty wiring. If you change enough parts enough times, chances are you'll eventually get the wiring right—even if you never knew that it was wrong in the first place.

Let's look on the troubleshooting chart and see what the problem really is when the gas valve won't open.

The "note" suggests reading through the procedure before beginning. This is called becoming familiar with the sequence of operation—a necessary part of troubleshooting. Walking though the chart here will probably take longer

QUICK & BASIC TROUBLESHOOTING

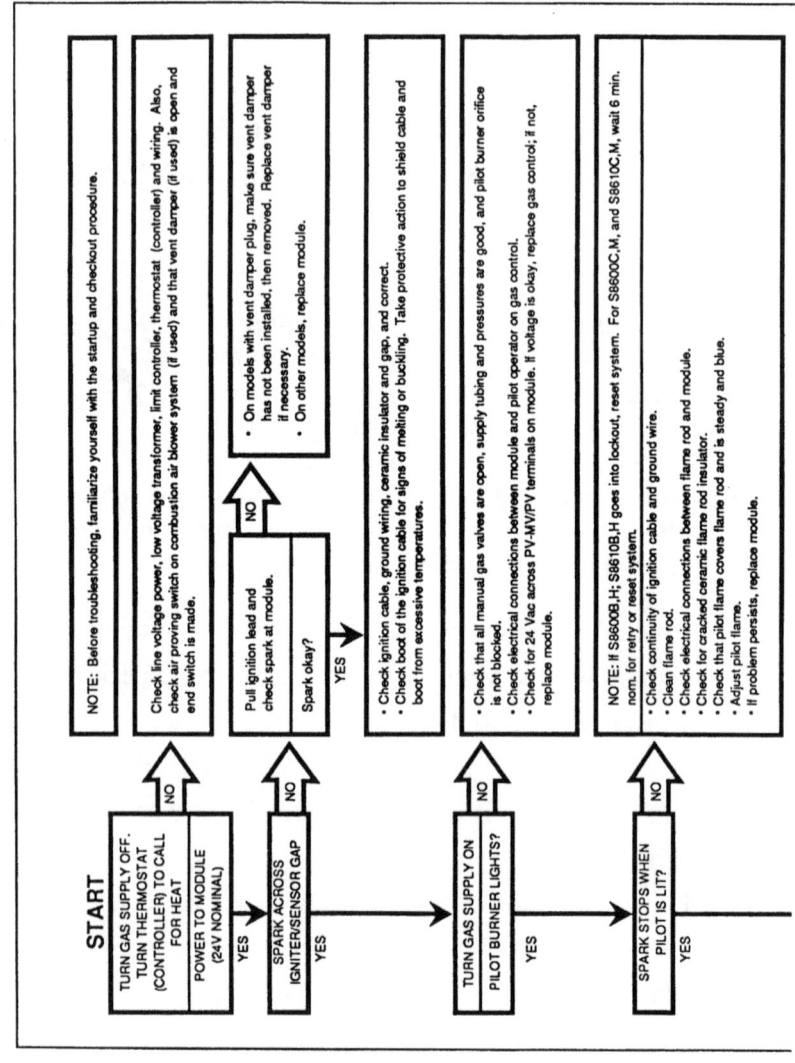

Troubleshooting the Complete Circuit

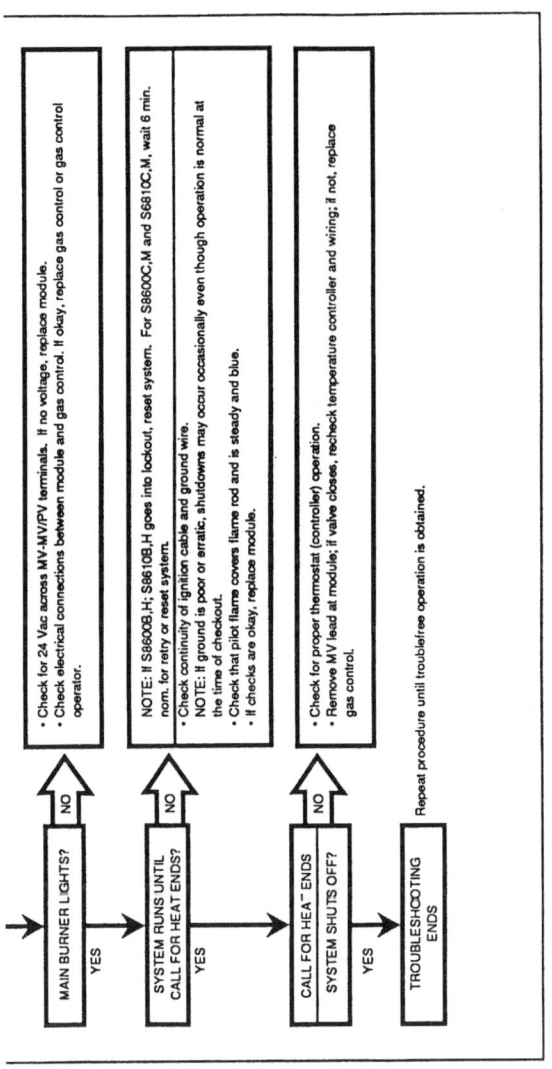

Figure 28. Troubleshooting chart

Compliments of Honeywell

QUICK & BASIC TROUBLESHOOTING

that doing the actual troubleshooting in the field. But a little time spent up front can save a lot later.

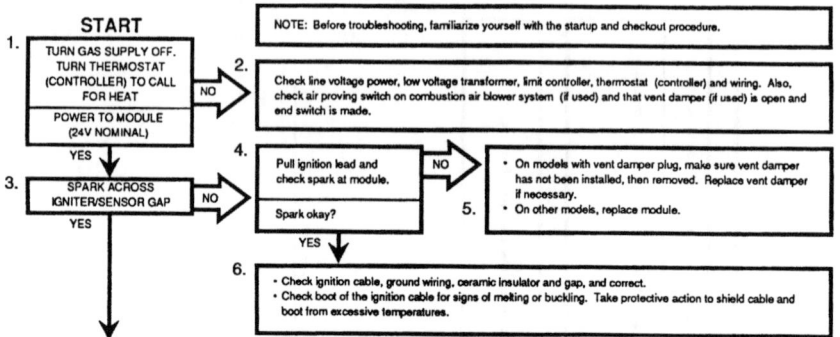

Figure 29. Top portion of troubleshooting chart

"Start" with box 1 in the upper left corner.

"Turn gas supply off." That's for your own protection.

"Turn thermostat (controller) to call for heat." Repeat after me: "A thermostat is just a switch, a thermostat is just a switch, a thermostat...." " Turning it to call for heat (whether you actually get warmth or not) turns on the switch.

In the same box, beneath the solid line, is "Power to module (24V nominal)." This means that as a result of turning the thermostat to call for heat, you should have 24V to the module. "Nominal" means close to 24V, say 23V to 27V.

More about "power to module." The module is the electronic box—the brains. The chart doesn't say where to measure. You'd have to look elsewhere in the technical literature. It turns out that the place to measure the voltage is module terminals labeled 24V and 24V GND.

Troubleshooting the Complete Circuit

Let's return to the chart. Coming out of the first box, you must choose "yes" there is at least 24V power to the module, which takes us to box 3, or "no" there is not, which takes us to box 2.

We'll follow "yes" first to box 3. If there is 24V to the module (and the system doesn't work), the next box asks if there is "spark across igniter/sensor gap." If you can answer yes, then you "turn gas supply on" (box 7) and determine if the "pilot burner lights." If yes, in the next box, can you observe that the "spark stops when pilot is lit?" (box 9). If yes, can you observe that the "main burner lights?" (box 11). If yes, does the "system run until the call for heat ends?" (box 13). If yes, when the "call for heat ends," does the "system shut off?" (box 15).

This last step is the same as "no problem." The system is working.

Let's return to the top of the chart, and select the first "no." That choice takes us to box 2.

This box is crammed full of actions. Be prepared to be here for awhile. In addition to the items listed, like in the movie *Casablanca* we might "round up the usual suspects": is the service switch "on," is the fuse good, and is the blower door switch closed?

Then we'll "Check line voltage power." This is an equivalent to "is it plugged in?" Line voltage means 120V coming into the building. With the meter set on VAC, touch one lead to each wire on the *primary* side of the transformer.

QUICK & BASIC TROUBLESHOOTING

Back to the chart, we're still in box 2, at "Check the limit controller." The limit controller is a switch put in the control circuit to make certain that the gas valve doesn't function if there's an unsafe condition. The most common limit controller is a high limit. If the temperature inside the equipment gets too high, the limit switch opens. That open switch assures that the equipment can't fire until it's cooled down to a safe temperature. Caution—a limit can be in either a 24V or 120V circuit.

Once you find the limit controller, you can do one of two things. You can use your meter to test to make sure the limit is *closed*. Use the continuity test or test for voltage drop, both described earlier.

If the limit tests as open, and if conditions are normal within the equipment (usually this means not too hot), perhaps the limit has failed. Check this by using a jumper across the two terminals.

If the original troubleshooting problem is resolved by placing this jumper across the limit, it means you have a bad limit. Figure out why the limit failed. They don't just wear out. Correct the situation, and replace the limit if necessary.

Let's continue back to box 2—"check thermostat." The thermostat (along with its subbase if there is one) is just a switch, just like the limit controller is a switch. Check the thermostat in exactly the same way as the limit controller above.

When you're working at the site of the equipment, you might not want to physically go the location of the thermostat. You can "jumper out" the thermostat by jumpering

Troubleshooting the Complete Circuit

across the two thermostat terminals on the module. When you use a jumper, think about what you're simulating. A jumper has the effect to taking out of the circuit whatever is between the two ends of the jumper. If the thermostat or the wire to it is the problem, the entire troubleshooting problem will be reduced when you jumper out the thermostat. In that case, replace the thermostat or the wiring.

Back to the top of the chart. We're back in box 2. We're still addressing the fact that we got a "no" to the question "is there power to the module?"

"Check wiring" is our next direction. We're looking for damaged or disconnected wires. Don't assume it doesn't happen. Use your hands to trace each wire and make sure it really goes from one terminal to another. Make absolutely certain that both ends are connected firmly to the correct terminals. Also look for nails or staples that may have penetrated a wire and caused a short.

If you can get the meter probes to both ends of the wire, the simplest test is to use the ohmmeter function for a continuity test. If the meter beeps, the wire is probably good.

Moving on within box 2, "Check air proving switch on combustion air blower system." Locate the air proving switch(es) and test just as you did the limit controller and thermostat.

Next, "Check that vent damper (if used) is open." This is a visual check.

Next, "Check that damper end switch is made." Check with your meter.

QUICK & BASIC TROUBLESHOOTING

By the end of box 2, you will have found what is preventing power to the module. If you haven't, start over again at the beginning of the box. It has to be here.

Tempting as it may be, don't fall for the idea of replacing the module. If there's no power to this module, there won't be power to its replacement, either

Let's assume that we found power to the module. That gives us a "yes" coming out of the box. We can move on to box 3, which asks, "Spark across igniter/sensor gap?" Be patient—sometimes there's a built-in delay. But let's say that our answer is "no."

Box 4 asks us to "Pull ignition lead at module." You should be able to draw a spark between the module terminal and ground using a jumper. Caution—this is high voltage. Do not hold the wire in your hand.

Let's choose "no" to spark okay. That takes us to box 5. "On models with vent damper plug, make sure vent damper has not been installed, then removed." This is a visual check. If there has been a vent damper plug removed, there's probably evidence of a damper. The box continues with "Replace vent damper if necessary." It concludes with "On other models, replace module."

But ask yourself, why did it fail? When replacing a module it's a good idea to put an ammeter or 3-amp circuit breaker in the "R" transformer run. The purpose is to watch for possible problem spots as loads, such as the gas valve, come on.

If there was ever a vent damper powered with the module, the module won't work without the damper. That's be-

Troubleshooting the Complete Circuit

cause powering through the damper plug blows a fuse inside the module. The purpose is safety—to make sure that there isn't a closed vent damper when the system fires.

Returning to box 4, let's select "yes," " the spark is OK." That puts us in box 6. This box is all about the ignition cable and the ceramic insulator.

So, if we can pull a spark from the module, but don't have a spark across the igniter/sensor gap, in box 6 we are asked first to "Check ignition cable." Think of a cable as a wire, and think continuity check. If there's not continuity, replace the cable. Also check it visually for cracks and signs that it's been overheated.

"Check ground wiring" comes next in box 6. This means that there has to be wire from "gnd" on the module to the burner. It's gotta be there. Add it if it's not there. Not having one is the number one reason for electronic ignition systems not working right.

"Check ceramic insulator" for cracks.

"Check gap" to make sure it's reasonable.

"Check boot of the ignition cable for signs of melting or buckling. Take protective action to shield cable and boot from excessive temperatures."

So, if you get to the end of box 6 and you haven't isolated the problem, try box 6 again. Moving on in the chart is tempting, but would be a mistake.

Let's return to box 3. This time let's answer "yes," there is a spark across the igniter/sensor gap. That takes us to box 7—"Turn gas supply on—pilot burner lights?" A "no"

QUICK & BASIC TROUBLESHOOTING

answer takes us to box 8. In box 8 we are first asked to "check that all manual gas valves are open." These valves would be in the gas line before the automatic gas valve. When the manual valve handle goes the same way as the gas line, the valve is open. When it's at right angles, the valve is closed. Also check the main gas valve to make sure it is turned on. (This one cost me a whole day once. It turned out that nothing was wrong except that the gas valve had somehow gotten turned off!)

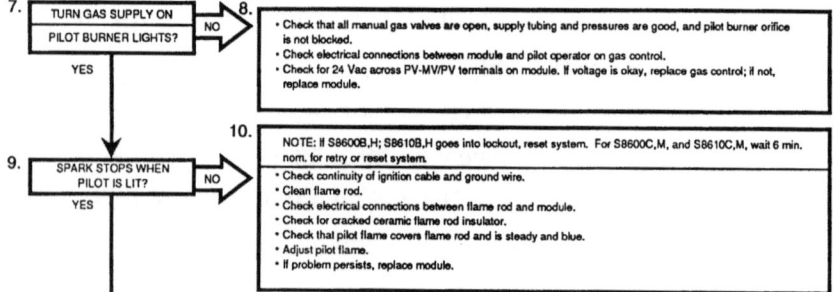

Figure 30. Middle portion of the troubleshooting chart

"Check supply tubing" is a visual check to be sure that tubing is intact and connected at both ends.

"Check that pressures are good" refers to gas pressure. Use a manometer to make sure there's at least 5" water column of gas pressure coming to the valve when all other gas appliances are running.

"Check that pilot burner orifice is not blocked." Maybe a spider decided to store a dead fly there.

Troubleshooting the Complete Circuit

"Check electrical connections between module and pilot operator on gas control." Here's one you'll probably hate. Do it meticulously anyway. *With your fingers,* not just with your eyes, actually trace each wire from the module. First gently tug on it to make sure it's firmly connected to the module. Then follow it to make certain it actually goes where it's intended. And make certain it's firmly attached on that end, too. (Note: On old valves TH-TR wasn't common. Putting MV/PV to it got pilot but no main burner.)

"Check for 24 Vac across PV-MV/PV terminals on module. If voltage is okay, replace gas control. If not, replace the module." PV (stands for pilot valve) and MV/PV (main valve/pilot valve) are terminals on the module. Simply set your multimeter to the lowest VAC (or V with a squiggle over it) setting. Touch a probe to each of these two terminals and see what reading you get. If there's at least 24V, the gas valve is bad, so replace it. If there isn't 24V, replace the module.

[Interesting but non-essential information. Here's how to think about the conclusions above of what to replace. The terminals PV and MV/PV are where power comes out of the module for the gas valve. If there's no power out, that means the module failed to provide it. If there is power out, the module is doing its job of providing the power. But the gas valve is failing to respond to the power.]

QUICK & BASIC TROUBLESHOOTING

Keep firmly in mind that these conclusions work *only* if you've followed the steps of the chart. Consider this as an example of how you can reach the wrong conclusion by skipping steps of starting in the middle of the chart.

Let's say you picked up the chart and your eyes fell first upon box 8, third bullet. So you check for 24Vac across the PV and MV/PV terminals. It's nice (and very misleading) to see the easy choice of either replace the module or the gas valve. The problem is that you haven't done important things like make certain there is power *into* the module or that there is gas available to the valve.

Let's return to box 7 and take the "yes" choice this time. That choice takes us to box 9 "Spark stops when pilot is lit?" Let's choose "no." That answer takes us to box 10.

Box 10 is the big one. Start with "Note: If [certain models] go into lockout, reset system." Lockout means that for safety reasons the module was designed to shut down if it failed to light the system. You can expect this type of module to say "100% lockout" on it. Reset means that you briefly remove power from the module in order to start it up again. Note continued: "For [certain models] wait 6 min. nom for retry or reset system." These models normally wait 6 minutes between each of their trials for ignition.

"Check continuity of ignition cable and ground wire." This looks like what we saw in box 6. It looks like repetition. But keep in mind that if during normal troubleshooting we had followed a path that got us to box 10, it would not have taken us through box 6.

"Clean flame rod." Use emery cloth or sandpaper to lightly clean the flame rod. It may have become corroded or coated with contaminants such as soap film from the laundry area.

"Check electrical connections between flame rod and module." Like you've seen before, this means using your fingers to make sure that both ends of the wire are firmly connected where they're supposed to be.

"Check for cracked ceramic flame rod insulator." You saw this before in box 6.

"Check that pilot flame covers flame rod and is steady and blue." The flame must always be on the flame rod. If it wavers or misses the rod, incorrect information is sent to the module. Here's the principle. The spark continues until the module receives a signal that the pilot is lit. The signal is created by the pilot constantly covering the flame rod. If it's on the rod only part of the time, the module thinks the pilot isn't really there, so it keeps the spark going.

"Adjust pilot flame" if it doesn't steadily cover the flame rod.

"If problem persists, replace module." The conclusion to replace the module comes only after we've determined that there's reason to believe that the module is receiving all the information it needs, but for some reason is unable to process the information.

Now let's go back to box 9 and this time choose "yes," for "the spark stops when pilot is lit." This moves us to box 11, which asks "main burner lights?"

QUICK & BASIC TROUBLESHOOTING

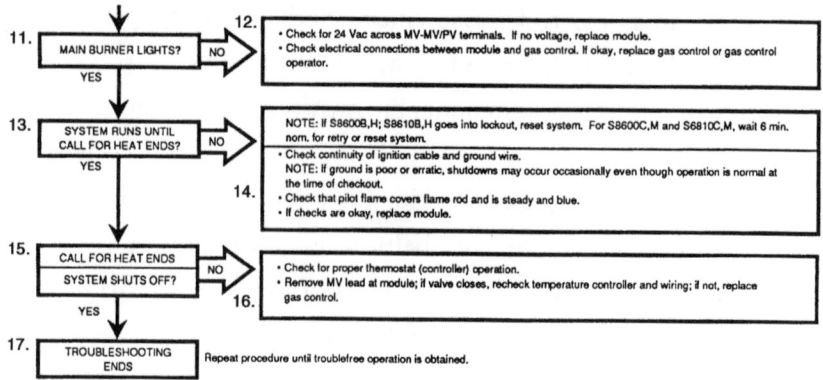

Figure 31. Bottom portion of the troubleshooting chart

Choosing "no," to "main burner lights" takes us to box 12. There we see "Check for 24 ac across MV-MV/PV terminals." This is not the same set of terminals as in box 8. This time we're looking at MV terminals, which is the main valve. This makes sense, because back in box 11, the main burner is in question. The testing process is the same as in box 8, however.

Here's one of those places where the troubleshooting chart takes us to a different conclusion that our own "logic" might take us. For many of us, the logic is this: if the burner won't light, it must be because it's not getting gas. The gas valve controls gas. So we conclude (in error) that it must be a faulty gas valve. This conclusion is supported by the fact that there is pilot flame to light the gas, if there were any.

What the troubleshooting chart is showing us is that if there's no voltage coming out of the module, it's the module that has failed to perform.

The second bullet in box 12 gets us to the gas valve. "Check electrical connections between module and gas con-

Troubleshooting the Complete Circuit

trol. If okay, replace gas control." So, if there is voltage coming from the module, and the wiring connections are good, there will be voltage going to the gas valve. Use your meter set on VAC to check for 24V between the MV and MV/PV terminals on the gas valve. If there's 24V going into the gas valve and the valve doesn't open, you have a bad valve.

Let's move back to box 11 and take the "yes" option. That takes us to box 13 "System runs until call for heat ends?" The "no" answer takes us to box 14. The note in box 14 is identical to that in box 10. If you were troubleshooting in real life, the chart wouldn't land you in both of these boxes, so you wouldn't be dealing with repetition.

Moving on in box 14, "Check continuity of ignition cable and ground wire." This is the same as in boxes 6 and 10. "Note: If ground is poor or erratic, shutdowns may occur occasionally even though operation is normal at the time of checkout." This is the same ground that we talked about in box 6. It's a ground between the module and the burner. It's important because it insures that the module receives correct information about the presence of flame.

"Check that pilot flame covers flame rod and is steady and blue." This too is the same as box 10.

"If checks are okay, replace module." So, you can see that box 14 is the same steps as box 10. But the reason you're checking is different. The module controls several different functions.

Returning to box 13, let's choose the "yes" option. That moves us to box 15, "[When] call for heat ends, system shuts off?"

QUICK & BASIC TROUBLESHOOTING

A "no" in box 15 takes us to box 16, "Check for proper thermostat (controller) operation." Since a thermostat is a switch, go back to the explanation of box 2. We're going to use the multimeter to check for continuity in the switch. Once again, in real-life troubleshooting, this wouldn't be repetition, because the chart wouldn't take us to both box 2 and 16.

This time we're checking to make sure the thermostat switch is open (drawbridge is open, switch is off) at the end of the call for heat.

The second part of box 16 is "Remove MV (main valve) lead at the module; if valve closes, recheck temperature controller and wiring." This means that if you remove power from the valve and the valve closes, the valve is okay. The problem has to be somewhere else. If the valve stays open after you remove power, (and this would be extremely unlikely) the valve has stuck open.

Now, back to box 15 for the "yes" option, we are taken to box 17 "Troubleshooting ends." If the problem wasn't resolved, start over again at the top of the chart. (It could be a long day!)

Troubleshooting Using a Ladder Diagram

A ladder diagram is a type of equipment diagram. Equipment diagrams are provided by furnace and boiler manufacturers. A ladder diagram is intended for troubleshooting, after a system has been installed. When a piece of equipment doesn't work, it's very useful to know where its power is coming from and what switches control it. A ladder diagram shows us the logic of how things work together.

Troubleshooting the Complete Circuit

The other type of equipment diagram is for original installation. It shows how things are to be wired together. It goes by several different names—connection diagram, hook-up diagram, or schematic diagram. Both this diagram and the ladder diagram often come side-by-side on one piece of paper, so watch out for labels.

A ladder diagram gets its name from its basic framework.

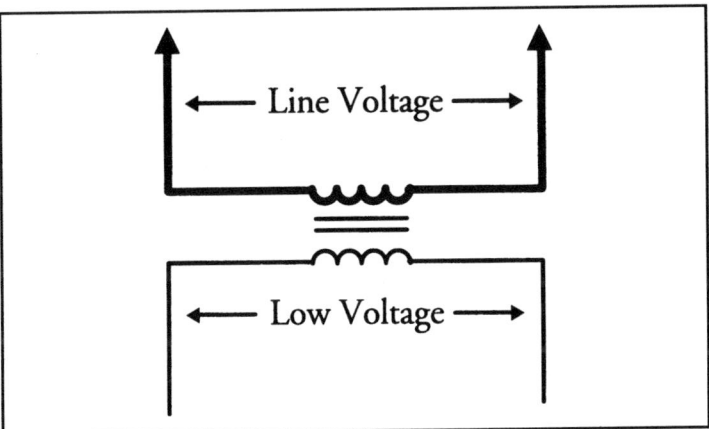

Figure 32. Ladder diagram framework

Notice a couple familiar things about the ladder diagram. In the middle of the ladder diagram is the symbol for the transformer. On the 120V side of the transformer (top of the drawing) are heavy lines, for line voltage. On the 24V side of the transformer, the lines are light.

Using this basic framework, we're going to add *steps* to the ladder. There will be a step for each load. On that same step will be the switch or switches that control that load.

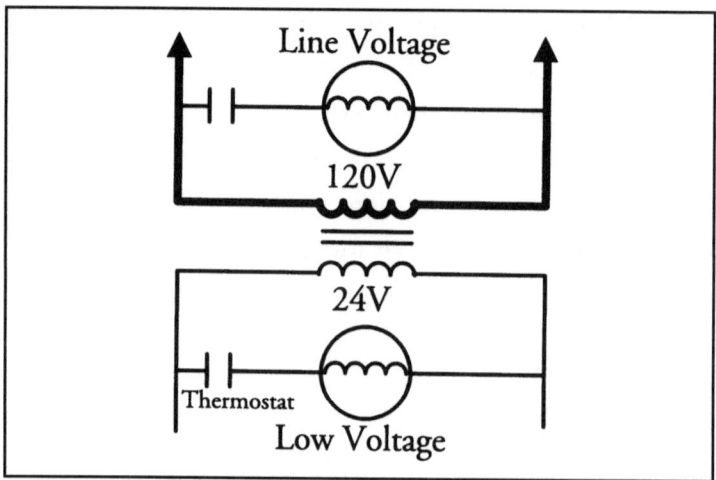

Figure 33. Ladder diagram with steps

The load and switch on the step, plus power supply, always make a complete circuit. In the diagram above, you can see that the low voltage load is controlled by the thermostat. It is powered by the 24V side of the transformer.

In the same diagram above, the line voltage step shows the line voltage load, controlled by a switch, and powered by line voltage.

We can add loads to either the line voltage or low voltage side by adding steps. Each step has one load and at least one switch.

Notice that the steps of the ladder make lines that are parallel with each other. Interestingly, the loads on each of these parallel lines are parallel loads, so long as they are within the same voltage. Recall that a definition of parallel loads is that they share a power supply, but are wired to op-

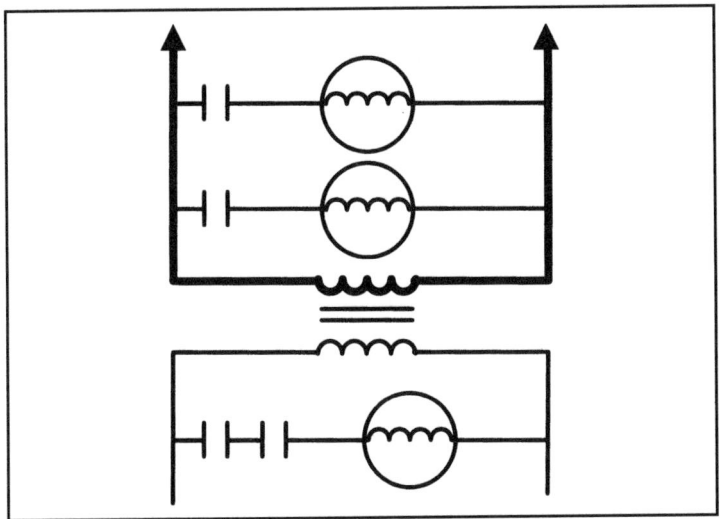

Figure 34. Ladder diagram with additional steps

erate independently of each other. Use the ladder diagram above to confirm this for yourself.

In the diagram above, the low voltage load is controlled by two switches. There can be multiple switches for a load. These switches are in series. Recall that "series" means that both switches must be closed for the circuit to work. A ladder diagram is particularly useful for identifying all the switches. You can imagine how frustrating troubleshooting would be if you weren't aware every switch that could keep a load turned off. Remember, when switches are in series, such as safety switches, it only takes one open (turned off) switch to disable the load.

[Interesting but non-essential information. Think of switches in series as being like a safety committee. Each committee member (switch) is responsible for a particular safety situation. The high limit makes sure the equipment

QUICK & BASIC TROUBLESHOOTING

doesn't get too hot. The flame roll-out switch assures that flame isn't where it doesn't belong. If any one of the committee senses the unsafe situation it's watching for, it opens (turns off) its switch. It votes "no." One "no" vote is all it takes to prevent lighting the equipment.

These switches are called "limits," short for limit switches. They limit the conditions under which the equipment can operate. When there are lots of limit switches, it means lots of conditions are being monitored for safety.

It's possible to "jumper out" a limit switch. Don't do that except for troubleshooting purposes. Taking out a limit is like taking out the traffic light in a busy intersection—you'd be creating an accident waiting to happen.]

Here's a simplified hook-up diagram (a hook-up diagram is used for original installation of equipment).

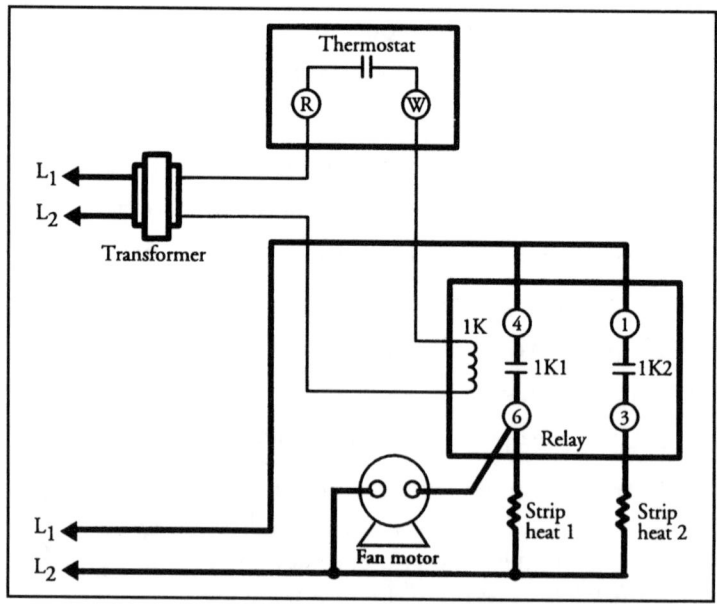

Figure 35a. Hook-up diagram

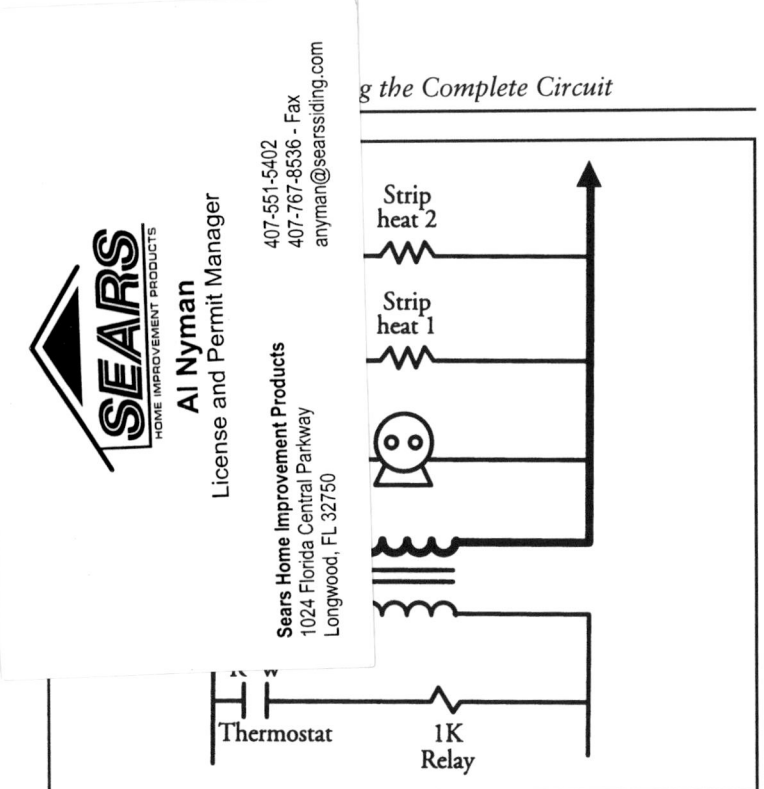

Figure 35b. Ladder diagram

Here is the ladder diagram of that same circuit.

In the hook-up diagram there is one low voltage load. So in the ladder diagram we would expect to find one step on the 24V side of the transformer. On that step we find the relay coil (labeled 1K) as the load, and the thermostat as the switch. With the 24V side of the transformer as the power supply, we have a complete circuit.

There are three line voltage loads in the hook-up diagram. We would expect to find three steps on the line voltage side of the ladder diagram. There is an individual step for the motor, for strip heat 1, and for strip heat 2. This might be electric baseboard heat. On the step with each

QUICK & BASIC TROUBLESHOOTING

load is the switch that controls it. Each load has separate access to the same line voltage power supply. We can see that the loads are in parallel with each other.

Let's look at the step with the motor as the load. What does the ladder diagram tell us about the switch? Because the switch is labeled 1K1, we know this switch is part of a relay. What would it take to close this switch? We would need to power the relay coil. When there is a relay switch, we can expect to find the relay coil in a circuit of another voltage. Look in the low voltage circuit for the relay coil.

We can see that to power the motor we must close switch 1K1. The only way to close that switch is to power relay coil 1K. Coil 1K is in the 24V circuit. It is powered only when the thermostat is closed. We can see that the low voltage thermostat controls the line voltage motor.

Moving back to the line voltage steps, let's see what powers strip heat 1. Puzzling as it may seem, we see that the switch controlling strip heat 1 is also relay switch 1K1.

And returning to the third step in line voltage, we find strip heat 2 controlled by relay switch 1K2. Once we get used to looking at a ladder diagram, we can see at a glance that relay switches control all three line voltage loads.

Now we're ready to look at an actual diagram that comes with a piece of equipment. First notice the headings along the top. Here's your clue that the page actually contains two separate diagrams, side-by-side. On the left side of the page is the schematic wiring diagram (it could also be called a connection diagram or hook-up diagram). On

Troubleshooting the Complete Circuit

the right side of the page is the ladder wiring diagram. That's the one we want to use for troubleshooting.

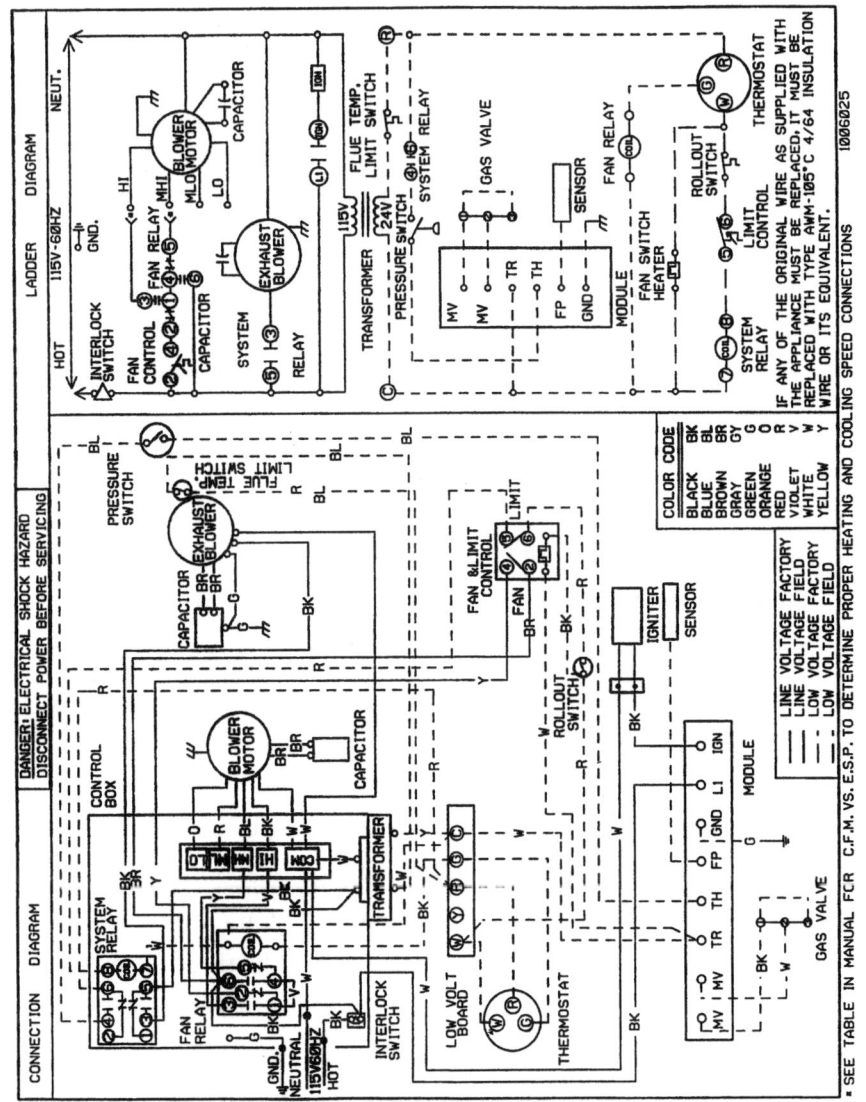

Figure 36. Equipment diagram

Diagram compliments of Intercity Products

QUICK & BASIC TROUBLESHOOTING

You can very quickly identify the loads—there is one load per *step* on the ladder. You also can see what switch controls each load. Perhaps most important of all, you can see if there is more than one switch controlling a load.

Let's look closely at this ladder diagram. First, find the transformer near the middle. Notice that the line voltage side (115V) is drawn in solid lines, and the low voltage side (24V) is shown in dotted lines.

On the low voltage side, on the bottom step of the ladder diagram, we find a load labeled *coil* and *system relay*. Because of what we know about relays, we can assume that when this relay coil is powered, one or more switches will change from open to closed, or vice versa. Those switches will be in circuits different from the one that this coil is in. Look elsewhere in the diagram for a *system relay*. It's also labeled with terminals 5 and 3. On that step is a load labeled *exhaust blower*. To power the exhaust blower, we must close the system relay switch.

Still reviewing what we know about relays, what does it take to close this system relay switch? We need to power the system relay coil. And how will we power that coil? Let's look at that coil again (low voltage side, bottom step). We need to close all the switches in that circuit. How many switches are there? At first there seems to be three. There's the *limit control*, the *rollout switch* and the *thermostat*. The first two are already closed. A call for heat will close the thermostat switch between terminals W and R. Is that all the switches? Remember, one of the most powerful

Troubleshooting the Complete Circuit

functions of the ladder diagram is that it lets us identify all the switches that control a load.

There's one more switch. Let's trace the complete circuit for the system relay coil. Starting at the coil, move to the right through the three switches. Continue to trace up toward the transformer (through terminal R) and, oops, there's one more switch—the *flue temp. limit switch*. It's closed. Trace on through the low voltage side of the transformer, and complete the circuit back at the system relay coil. We have a complete circuit. And we know now that there are *four* switches controlling the relay coil.

To look at all this information another way, ask the question, "What does it take to bring on the exhaust blower?" Think: power supply, switch, and load. The blower's power supply is line voltage. The switch is the system relay switch. The load is the exhaust blower. The system relay switch is open—what does it take to close it?

To close the relay switch, we must power the relay coil associated with that switch. We would expect the relay coil to be in a different circuit. We could also expect it to be in a low voltage circuit. We know from looking at the diagram that the coil is on the bottom step of the ladder diagram.

We are still tracing backwards to answer the question, "What does it take to power the exhaust blower?" What does it take to power the system relay coil? We must have 24V from the transformer. And we must have all four switches closed. All are normally closed except the thermostat. We need a call for heat to close the thermostat.

QUICK & BASIC TROUBLESHOOTING

All of this explanation is the long answer. The short answer—if everything is working right—is that a call for heat brings on the exhaust blower. If everything isn't working, the clues to making it right are in the ladder diagram.

Conclusion

Troubleshooting is a process. Like starting a car, you have to start at the beginning of the process, follow all the steps and not skip around. You can troubleshoot individual components or complete circuits. It's useful to troubleshoot with a troubleshooting chart or ladder diagram. But if a chart isn't available, you can use a troubleshooting process to find and fix the problem. Have patience, take your time, and practice.

Final Test

Answer true or false.
1. Troubleshooting can be done without a meter.
2. With a troubleshooting chart, start where your instincts say is best.
3. Jumpering the switch is a good troubleshooting technique.
4. Jumpering the load is a good troubleshooting technique.
5. To measure continuity, set your meter on ohms.
6. Continuity across a switch means it's open.
7. If the transformer secondary shows 24V, you need to check the primary.
8. Too big a load causes the transformer primary to burn out.
9. A ladder diagram shows how to hook up the original installation.
10. You can count on each circuit having only one switch.

Index

amp rating, 39–41
amperage, 33
anticipator, 56

bug-on-a-rope analogy, 23

circuits
    basic, 21–23
    series, 24
    parallel, 25
connectors, 24
continuity, 31, 42

daisy chain, 48
diagrams
    equipment, 87
    hook-up, 84
    ladder, 80–83, 85
    schematic, 84

ground, 38

home run, 48
hop scotch, 48

intermittent, 58

jumper, 54

leap frog, 48
loads, 21, 24–25, 44

meters
    buying, 26–27
    ammeter, 34–35
    multimeter, 28
    ohmmeter, 32
    voltmeter, 30

panels, 46
parallel circuits see "circuits"
power supply, 21, 36–37

relays, 45–46
resistance, 31

series circuits see "circuits"
switches, 21, 41–44

transformer, 37–40
troubleshooting chart, 62–80

VA rating, 39–41
voltage, 29–30
voltage drop, 42–43, 54

Order Information

To order copies of

Quick & Basic Electricity
Quick & Basic Hydronic Controls
Quick & Basic Troubleshooting

Mail this form to:
P.I.G. Press
759 E. Phillips Dr. S.
Littleton, CO 80122-2873

Enclose $20.00 US funds for each book plus $3.95 for shipping and handling. Price is subject to change.

Indicate which book(s) you wish to order and the quantity of each.

Your name _____
Your address _____
_____
Phone (if we have a question) _____

Questions? Call P.I.G. Press at:
303/795-2679
or fax 303/795-9350

Thanks!